Gadgets to God

Relections on our Changing Relationship with Technology

David Wortley

Copyright © 2012 David Wortley

The moral right of the author has been asserted.

Apart from any fair dealing for the purposes of research or private study, or criticism or review, as permitted under the Copyright, Designs and Patents Act 1988, this publication may only be reproduced, stored or transmitted, in any form or by any means, with the prior permission in writing of the publishers, or in the case of reprographic reproduction in accordance with the terms of licences issued by the Copyright Licensing Agency. Enquiries concerning reproduction outside those terms should be sent to the publishers.

Matador
9 Priory Busines Park
Wistow Road, Kibworth
Leicester LE8 0RX, UK
Tel: 0116 2792299
Email: books@troubador.co.uk
Web: www.troubador.co.uk/matador

ISBN 9781780881546

Typeset in 11pt Times by Troubador Publishing Ltd, Leicester, UK
Printed and bound in Great Britain by TJ International Ltd, Padstow, Cornwall

Matador is an imprint of Troubador Publishing Ltd

CONTENTS

	Preface	ix
1.	Is there anybody out there?	1
2.	My name is Ivor, I'm an Engine Driver	10
3.	Just got to get a Message to you	56
4.	Leapfrogging with Ghengis Khan on Level Five	117
5.	Oh what a Picture, what a Photograph!	157
6.	Video killed the Radio Star	191
7.	When will we ever Learn?	235
8.	See me, Feel me, Touch me, Heal me	267
9.	It's only a Game	304
10.	The End of the Beginning or the Beginning of the End?	335
	Index	359

PREFACE

This book is a journey through the gadgets and technologies that have influenced my life and are about to bring almost unimaginable changes to our world. Each of the chapters is devoted to an individual communications technology from the transport technologies of bicycles, scooters, motor bikes, cars, buses, trains and planes of chapter two to the immersive technologies of video games, virtual worlds and social networks that now engage our discretionary time, attention and income. The chapters have been organised in the chronological order in which they influenced my career and my life and, on reflection, are in increasing order of importance and significance in determining the future of everyone on this planet. The span of my lifetime has seen a massive transformation in our use of these technologies as they now converge within powerful new devices and applications. With this transformation in usage has come a profound change in our relationship with technology which I believe will create unprecedented challenges and opportunities that we may struggle to manage.

The message behind this autobiographical account of my own and mankind's relationship with machines and technology in the second half of the twentieth century and first decade of the twenty first is quite simple. My generation of baby boomers has lived through an unprecedented period in our history. My life began in an era when human beings used their intelligence and senses to get the best out of machines and technology and will end in an age in which machines and technology will be using all their intelligence and senses to get the best out of human beings. Where once we were masters we may now become slaves. The consequences of this shift in power are profound, unpredictable and potentially either disastrous or liberating for the future of mankind. The choices we are making today about how, where and when we use the technologies described in this book could well destroy civilisation and the beautiful planet we live on. The world I have spent my life in is a metaphorical "Garden of Eden" yet I fear that the new God we are all

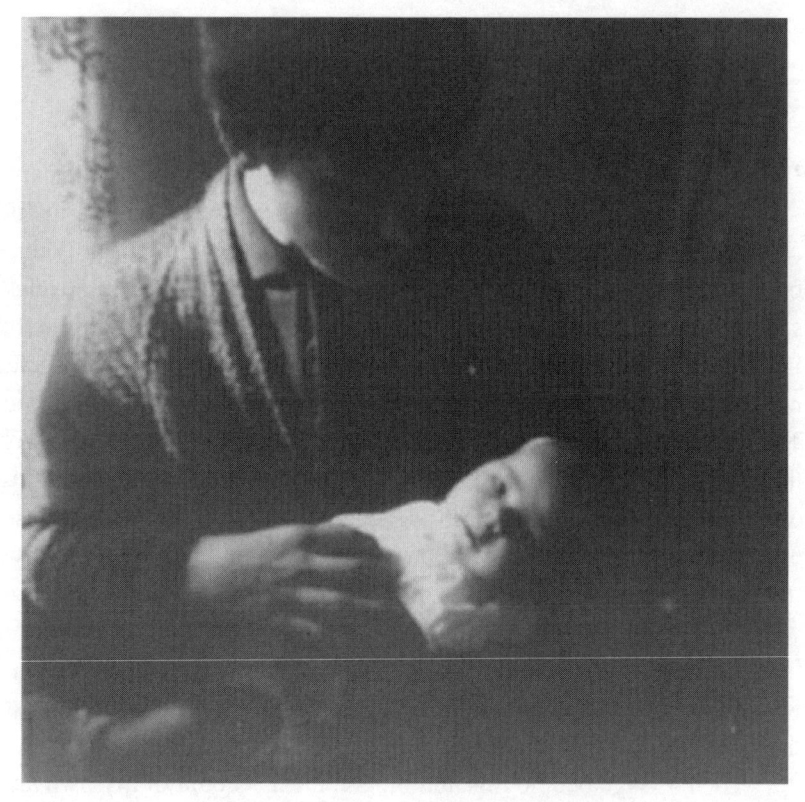

My son and I before his adoption in 1967

choosing to create in our own image and empower with all the knowledge of good and evil may well cast us out into an age of darkness and conflict. The stories of the Bible's Old Testament may not have been about the past but may well be predicting the future.

My wish is that this book will hold something of value for everyone with an interest in what the future may hold. I hope that the "Digital Natives" who were born into the "Information Society" of the internet will learn what the world was like without computers, broadband, mobile phones, video games and social networking web sites and use this knowledge to harness these powerful tools for the good of mankind. For the "Digital Immigrants" of my own baby boomer generation who have taken instructions from a three year old grandchild on how to use a video recorder control, I hope that my memories will provoke nostalgic reminiscences of a bygone world yet act as a wake-up call to alert them to what we may all be cosily sleep-walking into.

I am indebted to my good friend Sally Ann Moore for reminding me of the "boiling frog" anecdotal story (as used by Al Gore in his film "An Inconvenient Truth") in which a frog placed in boiling water will jump out but if the water is heated slowly enough it will stay in until it is boiled alive. Today we all may be that frog but the rate of heating is now accelerating rapidly and I hope we still have time to jump out the saucepan before the lid is sealed on our fate.

The book is dedicated to my mother and father and my grandparents who played such an important part in making me who I am, to my son whom I haven't seen since he was placed for adoption in 1967 and to the technologies that have so shaped my life.

CHAPTER ONE
Is there anybody out there?

"The reasonable man adapts himself to the world; the unreasonable one persists in trying to adapt the world to himself. Therefore all progress depends on the unreasonable man."

George Bernard Shaw
Man and Superman (1903) "Maxims for Revolutionists"
Irish dramatist & socialist (1856 – 1950)

We human beings are virtually unique in the animal kingdom through our ability to individually and collectively shape the environment in which we live. We have an inbuilt desire to challenge the physical constraints of our universe and become masters of our own destiny. We are born into a world which has its own highly evolved ecology and seasons that sustain both humans and all other living beings and yet we strive for more. We gain comfort from the natural rhythm and laws of nature whose awesome complexity helps us believe in a powerful deity beyond our comprehension yet the vast majority of us are not willing to accept the boundaries and limitations that nature has provided. Blind faith in our existence and acceptance of the hand of fate are not enough for many of us as we strive for better understanding and control over the forces which shape our future. The advances that mankind has made in the span of my lifetime to bring control of these forces into the hands of individual citizens are unprecedented in our history. This chapter charts the start of that journey.

B.C. – Before Computers

It seems incredible to me now that, as a "baby boomer" born four years after the end of the Second World War, I came into a world where my family home had no access to any of the communications technologies that have been such an important part of my life and are so taken for granted by today's younger "digital native" generation. Our family home had no telephone, no car, no television, no high fidelity entertainment system and definitely no computer. The street lighting was still powered by gas in some parts, we had no bathroom, no central heating and our only means of transport were bicycles or public transport. Foreign travel for the ordinary man in the street was unheard of and yet we lived less than one mile from the centre of a prosperous market town.

I was born in the Lincolnshire town of Boston to a father who was a carpenter and joiner and a grand-father who was a gang foreman on the railways. My very early years were spent living with my parents and grand-parents in the family home in Boston before my mum and dad eventually got a council house of their own in the village of Wyberton – just south of Boston and within easy sight of the main railway line from Grimsby to London Kings Cross. Every morning at ten past eight I would hear the whistle of the Kings Cross – Grimsby – Cleethorpe express as it left Boston and I would run to the landing window to catch sight of a magnificent Britannia Class Locomotive as it came into view from behind the stacks of telegraph poles in the timber yard that stood at the bottom of our garden. Billowing smoke and with exhaust chattering, these engines would haul maroon carriages filled with unknown people bound for the unknown and unseen destinations that so excited my imagination. Regular Britannia Class locos used on this route were "Robin Hood" and "Oliver Cromwell" and they became almost like friends to me over my early years and, as I became older, they would carry me too to Peterborough and London.

It was my grand-father Sydney Plant Langford otherwise known as "Pop" who helped foster my love of engines and travel. Railwaymen in those days had concessions which allowed them free travel for the family across the whole railway network. My grand-parents would take me from a very early age all over the country at the weekends, mostly to cities where Pop and I would watch a football match whilst grandma did some shopping. In my bedroom at home, my dad had made me a wooden toy box which contained one of my most prized possessions, an engineering drawing of the rail network designed for the wood yard at the end of our garden. I used to study it and imagine that one day I might be a railway engineer. Amongst my earliest memories as a very small child was being given a present of a toy train and how I sobbed because it wasn't the right colour!

Once we had moved to Wyberton, close to the railway line, I would spend many hours by the crossing gate on Tytton Lane East watching the gates manually closed by Mrs Skinner, the resident crossing gate keeper who lived in the house next to the level crossing. I would crouch down and listen to the hum of the rails as the unseen distant trains would begin to come into view with freight hauled by clanking old WD (War Department) engines, expresses fronted by Britannia Class locos or more local passenger trains pulled by B1 Pacifics like "Mayflower" whose nameplate now resides in a Boston museum near to where the Pilgrim Fathers were imprisoned before making their historic voyage to the "New World".

Throughout history mankind's progress has been marked by quantum leaps in communications technology and man's harnessing of forces more powerful than his own to shape the world we live in today. We constantly seek to challenge the physical constraints of space, time and state. The power of animals like horses, oxen and elephants has been harnessed for centuries to achieve what man could not accomplish by his own strength and capabilities. To travel faster, carry heavier loads and work more productively, mankind has set animals to work since the dawn of civilisation. Could it be then that our attachment to animals as living creatures purposed to work in companion with their human masters and blessed with the same physical and mental frailties as human beings means that our affection for machines is a natural legacy of our relationship with these domesticated creatures?

It was through steam trains and the railways that my relationship with machines and technology began. It was the steam trains that enabled me to challenge my personal constraints of time and space and take me to new worlds and experiences in a time and at a distance beyond my own capabilities. My relationship with machines and technology was symptomatic of a belief that I had the ability to get something highly unique and personal from the machines and technology I used. Indeed, in the era into which I was born, the performance of machines and technology was very much dependent on the skill, experience and intelligence of the operator. This relationship with machines that we all feel to varying degrees may be the reason why these mechanical devices were often referred to as feminine and given female names, either as a sign of affection or in recognition of their unpredictability and mysteriousness. It is a curious thing that the most cherished machines are referred to as "she". Cars, ships and aircraft are invariably given female names and yet their mechanical nature could not be further from the essence of the female personality.

One exception to this generality was my Uncle Bill's car which, although it was often referred to as "old girl", was nicknamed "The Bomb". My Uncle Bill and Aunt Iris lived less than a hundred yards away on our street and were the envy of most families in Yarborough Road because they were the first household to have a car and the first in the street to get a television – a black and white model purchased, like so many others, to coincide with Queen Elizabeth's coronation in 1953. It just seems incredible to think that in my lifetime we have gone from having no home entertainment systems (apart from radiograms and gramophone record players) to being able to watch full colour high fidelity movies on portable devices and mobile phones.

So it was that every Tuesday night I would have tea with Bill and Iris and my two cousins Mick and Steve. Here I would share the twin delights of my Aunt's sponge cake and an episode of Popeye on the newfangled TV. This pleasure was counterbalanced by the less enjoyable shared bath with my cousins. In those days the water was heated in the winter by the coal fire and in the summer by using a gas-fired "copper" in the outhouse and then carrying in jugs of hot water to fill the bath with inevitably lukewarm water. At least it was some advance on the old tin bath my grand-parents had to use in their living room.

My fondness and relationship with machines and technology began with steam locomotives on the old LNER (London and North Eastern Railway), the steam driven cranes that worked in the wood yard, my uncle's car, our battery driven radio with mysterious and largely forgotten radio station names like Hilversum on its dial, the Lancaster and Spitfire aircraft my dad described to me from his wartime days and our own black and white television with its glowing valves and propensity for breaking down. All these machines and technologies I remember with great fondness and affection and I thought of them as personalities in their own right, serving me not as lifeless inanimate objects but as friends who would take me to new worlds and experiences at my bidding.

My grandmother had a wind-up clockwork driven gramophone record player with a collection of 78 rpm records and I remember to this day the constant playing and re-playing of my favourite record – "The flight of the bumble bee" by Spike Jones and the City Slickers. These days it would be regarded as a novelty record because it began with a fairly serious attempt at a piece of classical music interrupted by a sneeze and then disintegrating into gales of laughter from the orchestra as they tried unsuccessfully to restart the piece. I never could listen to this record without bursting into laughter myself.

My uncle's car (The Bomb) was our family chariot of delight at the weekends and on family holidays when it was used to take the four adults and three children to such exotic places as Grimsby, Skegness, Corby Glen (for train spotting) and Hunstanton (for the annual family camping holiday). The Bomb was a Humber Snipe which, in the days before electronic ignition, had to be started with a crank handle and required a weekly service from my uncle and my dad before we could risk setting out. This involved cleaning her spark plugs, topping up oil and water and checking the tyres. Despite all this care and attention the old girl (The Bomb) would regularly break down on our trips and our journeys always featured spots where everyone in the car would salute in recognition of the site of a breakdown. Apart from one occasion when the old girl refused to restart on a trip back from Skegness, my dad and uncle always managed to coax her back into life after some inspection and tinkering under the bonnet.

My fascination with steam trains was however undoubtedly the strongest and I spent many hours train spotting on Peterborough station hoping to catch a glimpse of an A4 Gresley Pacific like "Mallard" thundering through the station with the Flying Scotsman to Edinburgh. I also enjoyed family trips in "The Bomb" to Corby Glen in Leicestershire where we would stand on a hump back bridge listening for the beat and chatter of the exhaust of the express passenger trains as they sped through the cutting close to where the world steam land speed record of 126 mph was set by "Mallard" many years before I was born. This affection for steam locomotives accounts for some of my earliest, strongest and fondest childhood memories such as the seaside specials to Skegness where rows and rows of carriages stood in the sidings with their locomotives from all over the UK resting before the journey home, smoke still gently drifting from their smoke stacks.

The sense that these locomotives were living beings made me feel that I was in their care, that they would use their power to speed me to my destination to the best of their ability. I still recall the excitement I felt when engineering works on the main line between Huntingdon and Peterborough caused my Kings Cross – Grimsby – Cleethorpes express to be diverted via Cambridge, March and Ely. I was hauled by "Robin Hood" on that occasion and recall hurtling through the night, watching sparks flying past the carriage window and imagining that this locomotive was working so hard to get me back home as quickly as possible. I never thought that its speed and power might have anything to do with the driver and fireman, it was just the locomotive itself straining every valve and piston like a trusted friend doing its utmost for me.

My first personal attempt at anything like engineering was with a toy Meccano set bought as a Christmas present. I had long drooled over the pages of the brochure and longed to be able to construct my own transporter bridge or steam crane but the fiddly nuts and bolts, lack of pieces and my own low boredom threshold proved too much of a barrier. My next venture was an Airfix kit for the Lancaster bomber which my dad used to work as ground crew on in the RAF in India during the war. I did have more success with this and built both a Lancaster and a Wellington bomber. I remember that they cost me six shillings each which amounted to a full day's work runner bean pulling in the summer fields near my home. The Wellington Bomber had a short life as the mischievous part of my personality persuaded me to conduct an experiment one November morning by inserting a firework into the Wellington, lighting the fuse and launching it from my bedroom window in the hope it would explode spectacularly over our front lawn. The experiment was a failure and resulted in the only smacked bottom hiding I ever had from my dad.

This interest in building model aircraft led to a "commission" from one of my school pals to build him a Spitfire which could really fly. He gave me the kit which had lots of balsa wood parts that had to be cut out from templates and assembled with Bostik glue to create a frame over which some kind of paper was glued to create the fuselage. The power came from an elastic band fixed to the propeller and wound up like clockwork before being launched for a predictably short flight. This enterprise was also short-lived as I had far more success at gluing my fingers together than assembling this rudimentary aircraft.

My only other memorable interest in science in these formative years was from a chemistry set which did succeed in generating impressively noxious fumes from some of my experiments before the base of the vacuum flask was re-purposed as a crystal ball complete with flashing lights controlled by a handle under the table as part of a charity fund raising fortune teller stall I ran for the scouts.

Since this book is about relationships and technology I must say something about the importance of place, community and relationships. Most of my early life was spent living in a council house at the end of the Yarborough Road cul de sac. At the end of our garden there was a ditch separating us from Calders and Grandidge wood yard. Like most people I have an attachment to my home town and the memories it holds for me. We knew all our neighbours – to the left of our house lived the Gilbert family. Mrs Gilbert had a voice that would challenge the loudest sergeant in the army when she would shout for her son, oddly but appropriately named Major Gilbert. I can almost hear her now calling "Major" down the street with all the power her mighty lungs could muster. I am convinced that she could be heard in the playing fields over a mile away. To the right of our house lived a family called the Hipkins – the father Tom was a lorry driver and I remember him taking my mum down to London on a trip once, down the old A1 via Biggleswade. Tom, who was a practical joker, once told me that tea could be poisonous and it had such an impression on my 10 year old brain that I have only drunk coffee since.

After the Hipkins had moved on, our next door neighbours were the Summerfields, a quiet couple who had 2 children. The husband worked on the railways as a fireman which made him something of a hero to me

Google Streetview of the Yarborough Road cul-de-sac as it is today. My old home is second from the left

as he described how cold it was on the footplate when the engine was running "tender first" (in reverse). Their story ended in tragedy when Mr Summerfield died on the operating table after a kidney operation in which both kidneys failed. Mrs Summerfield went on to remarry an older bachelor and after some years tried to murder him with a hammer. She was committed to a mental hospital and her new husband drowned himself in the fresh water cistern in their back yard. I believe that when she was released back home from Rauceby hospital that she also committed suicide in the same way.

Elsewhere in our street we had the Brammers, the Burdasses (much like Pop Larkin's family in the Darling Buds of May TV series), the Boothbys whose budgie I taught to say "Silly old Frank" whilst I was looking after it during their holiday, the Burgesses whose son Andrew was in the same year as me at school and went on to Cambridge University, the Scrupps, the Lennons, the Browns and my Uncle and Aunt, the Langfords. In the 1950s and 60s the notion of physical place was very important – it was the bedrock of communities who very often spent their whole lives in the same street. Many of those who lived in our street would never have gone much further than the 25 miles to Skegness and almost certainly never as far afield as London. I still love my home town and all the familiar places from my youth. The village of

Wyberton with its church, scout hut, village hall and local shop are all ingrained in me in ways which seem long forgotten as the following decades brought more mobility, less ties and a disintegration of the concept of a physical community. Wyberton and the town of Boston were defined by the local personalities and events of the time and it felt as if the village and the town actually "knew" me when I returned home. It is the changing importance of place which is now being re-shaped by technologies that I will refer to in my last two chapters.

The sense of community and the inter-dependence of neighbours and sources of authority were very strong in my early years. Every one of us was dependent in some way on the individuals and entities that shaped our lives. We relied on our relationships with other human beings and our faith in an all-powerful and beneficent God to develop our lives and shape our future. Machines and technology could only play a small part in helping to fulfil our aspirations and our dreams. It was the wisdom and experience of those we loved and respected that shaped our development and created opportunities for us to fulfil our potential, not machines or technology.

I was brought up in a Christian family and although we were never regular church goers, I have always had a faith in God as an unseen force behind the miracle of our existence. As a child I would pray every night as my parents had taught me and have always believed that some higher power is watching over me and helping to direct my life. Looking back now and reflecting on these early years with some degree of affection and sense of loss of a bygone age that I can never re-visit, it does seem that I was blessed with an entrepreneurial creative spirit and a low boredom threshold, factors which have played a large role in my love of and relationship with machines and technology throughout my life.

Let the journey begin……

CHAPTER TWO
My Name is Ivor, I'm an Engine Driver

Fulfilling a boyhood dream at the controls of Britannia Class Locomotive Oliver Cromwell

Since the dawn of time, mankind has sought to master the physical dimensions of the universe. Our own physical characteristics place constraints on our ability to manipulate space, time and state. We can only run at a certain speed, jump a certain distance or lift a certain weight. These constraints are simultaneously comforting yet challenging. They are comforting because they allow us to predict and plan for the future and are an essential part of a sustainable society that can operate within predictable norms. They are challenging because mankind has an inherent desire to break through the barriers of the physical world and, alone amongst all other living creatures, we have the intelligence both individually and collectively to find ways to extend our own capabilities. Early mankind developed tools to extend physical strength. We domesticated animals which had more power and speed than humans and for hundreds of years we have designed and built machines and infrastructures that that would bring about constant improvements in what any individual could achieve by their own effort.

All of the chapters in this book are about the machines and technologies we have developed to break down the physical constraints of our universe. Transport was probably the first communications technology to begin to challenge the dimensions of space and time. The use of horses not only extended the distance human beings could realistically travel but also reduced the time taken for any journey. The invention of the wheel allowed mankind to move heavier loads. Sailing boats conquered the oceans and protected the sailors from the harsh environment. The Romans built roads that allowed their soldiers to move swiftly across Europe and nations who mastered these transport technologies built empires and societies around them. It wasn't until the industrial age of the 19th century however that the world saw really significant improvements in our ability to conquer space, time and state. It was a little over a century between the first steam railway and the first man in space. The early steam railways saw a man walking in front of the locomotive with a red flag to prevent accidents and many scholars warned of the dangers of these "excessive speeds" beyond which man's body was designed to cope with. By 1938, the LNER Locomotive "Mallard" had set the world's speed record of 126 mph between Grantham and Peterborough, a record that stands today, and in 1969 mankind had made its first journey to another planetary body, the moon. Alongside all of these developments has been the consumerisation of travel technologies to make this increasing mastery of space and time accessible and affordable for everyone. This chapter charts my personal experience of these changes and marks the beginning of this book's journey to the opportunities and challenges of the 21st century.

Trains and Boats and Planes

The flat rural landscape of Holland in Lincolnshire was both a dream and a nightmare for the railway companies. For the engineers there was the luxury of straight line routes with no embankments or cuttings to consider but for the operators there were dozens and dozens of level crossings and the capital and human cost of gatehouses and gatekeepers at every level crossing. It was probably this combination of easy flat landscape coupled with the ambitions of local Boston business people to create faster routes to profitable markets and leverage Boston's thriving port activity that led to the rail route through Boston becoming the original Kings Cross to Edinburgh main line until it was relegated by the construction of the more direct route through Grantham and Retford on its way to Doncaster and York.

Communications technologies of all kinds have been the prime engines of disruptive change, increased trade, innovation and the development of civilisation. Those who have been masters of the communications technology of their era have built their empires around this mastery. Egyptian Hiroglyphics, ancient sailing ships, Roman Roads, the printing press, canals, railways, telegraphy, telephony, cinema, radio and television have all played a major part in dislocating changes in power and influence in both commercial and political arenas. With every quantum leap in communications technology there have been major advances in trade and the growth of civilisation but today, as I will outline in a later chapter, with the birth of the internet we have entered a period in our history which is radically different in nature to any of the previous quantum shifts.

Until the invention of the steam engine which launched the industrial revolution, the speed at which a man could travel and the volume of goods he could transport had changed very little over thousands of years. Roman roads had helped to create an empire in which armies and goods could be moved more easily from city to city and conflict to conflict but most trade was dominated by fleets of ships that bygone explorers, merchants and navies used to shape the ancient world. Canals also proved to be a successful internal transport infrastructure and a low cost and effective way of bringing goods and materials to market until the arrival of the railways.

As a regular international traveller myself, I continue to be amazed and extremely impressed by the explorers and adventurers of bygone centuries. These resourceful individuals would set sail for distant shores without knowledge of foreign languages or foreign currencies and would somehow negotiate trading deals that would bring exotic goods from far-flung places. Knowing the difficulties of modern day travel when large sections of the planet speak English very well, I can only marvel at their success.

One such individual who came to act as a major inspiration in my life was Thomas Cook, founder of the global travel empire and originator of the Travellers Cheque. I first discovered something of his story when I was living in Market Harborough in the 1990's. I auditioned for a part in a pro-am musical at the Leicester Haymarket Theatre to celebrate the anniversary of his very first travel excursion. It was during the three week run of this play called "Follow the Man from Cooks" that I first appreciated the extraordinary impact of his life not only on the travel industry but also on our daily lives, and the lessons we can still learn from him about how to innovate and shape the world we live in.

Thomas Cook was the son of a Methodist Preacher and born in Melbourne, Derbyshire. He got married and moved to Market

Thomas Cook – founder of the package holiday

Harborough where he plied his trade as a cabinet maker. He and his family lived in a tiny house in an alley called Quakers Yard in the centre of the town. It was a freak incident that changed his life and set in motion a series of events that led to the formation of one of the best known global travel brands from these humble and unlikely origins. He was attacked by a drunken man wielding the thigh bone of a horse and badly injured in the process. It was this experience that led him into the Temperance movement and a relationship with William Symington, a well-known business man in the town. They attended many temperance meetings in Market Harborough which, for a small market town, allegedly had one of the highest proportions of inns and ale house per head of population.

It was this passion for converting his fellow man away from the evils of alcohol abuse that led him one fateful day to set out walking to Leicester some 15 miles away for yet another temperance meeting. As he stopped to rest midway on his journey near the village of Kibworth, he had a blinding vision which suggested that he should "harness the power of steam" to the temperance movement. He suddenly realised that he could fulfil his passion to benefit his fellow man by opening up new horizons to them through travel to new places and new experiences. His reasoning was that if he could make it possible for the man in the street (or better still, the man in the pub) to journey to other towns, cities and even countries, their lives could be transformed and they would lose interest in alcohol. Sadly, his success in bringing tourism to the masses did little to curb an interest in alcohol and indeed it is arguable that he achieved almost exactly the opposite of what he intended !

Directly as a result of this vision, he organised the world's first package holiday trip by train from Leicester to Loughborough (around 12 miles) where about 600 people paid to enjoy a day out listening to temperance speeches and brass bands along with a picnic in the local park. Thomas Cook's success with this venture led him to organise many more trips, initially to major events such as the Great Exhibition in London and later on to many foreign lands and world tours where he often acted as the tour guide for his parties. Apart from admiring his enterprise and passion for the good of his fellow man, I learnt an enormous lesson about the power of communications technology and how it can be harnessed to change the world.

Thomas Cook used an emerging communications network to make a new and innovative service attractive, accessible and affordable. It is this combination of attractiveness, accessibility and affordability which is so relevant today as we seek to harness the power of modern communications infrastructures and technologies. He made his travel packages attractive by bringing dreams of new experiences into humdrum everyday lives. He made them accessible by tapping into a burgeoning and ubiquitous travel network and, importantly, removing any perceived barriers to travel through the package and tour guide concept and the invention of the Travellers cheque and he made them affordable by developing mass market appeal and negotiating bulk rates for his tour parties.

Thomas Cook was a typical social entrepreneur, not motivated by profit but passionate about his work for his fellow man. Some years after he moved his family home and his travel head quarters to the city of Leicester, his son John Mason Cook came into the business and rebelled against what he saw as his father's lack of business acumen. John Mason Cook eventually took over the business and removed his father from the company he had founded and Thomas Cook died a blind and broken man, devastated also by the death of his beloved daughter who was asphyxiated by a faulty gas heater in their bathroom.

It was just over 100 years after Thomas Cook's first railway excursion that I got introduced to the pleasures of travel to new places and experiences. One of my earliest memories of rail travel was being taken to London by my grand-parents to see the Harringay circus. I remember going down the wooden escalator at Kings Cross Station with the rows of globe lighting on stalks in the space between the up and down escalator and I remember seeing an injured man covered in a blanket at the bottom of the escalator.

Rail travel with my grand-parents took me to Sheffield where I watched Sheffield Wednesday play Nottingham Forest at Hillsborough whilst my grandma went shopping round British Home Stores and we travelled on the old trolley bus trams that used to run in the city. Such trips to cities to watch football matches and occasionally follow Boston United to away matches became a regular and much anticipated part of my life. My grand-father was a staunch but often critical supporter of our home team to the extent that when he had to work on home match days he used to still send his entrance money to the club, much to the disgust and dismay of my grandma, especially in my early years when Ration Books were still in use for certain items.

So it was that the railways became, as Thomas Cook had intended, a gateway to new worlds of unimaginable possibilities. Boston was a thriving railway town with its own engine sheds and as well as the direct north – south route from Grimsby to London Kings Cross which also profited from the early morning "fish train" from Grimsby to London. There were also links to Boston Docks from which timber was a regular freight, to Skegness via the famous Firsby junction curve, to Sleaford and Grantham for connections to Nottingham, Derby and Sheffield, direct lines to Horncastle and to Woodhall Spa and a line to Lincoln which ran for a great part along the bank of the river Witham.

The line to the docks had to cross the river Haven (the salt-water estuary of the Witham) via a swing bridge which was necessary because of the fishing smacks that needed to tie up and deliver their hauls on that part of the river. This swing bridge used to be a useful short cut on my way to Grammar school and I always looked forward to seeing one of the little shunting engines hauling bogie bolster wagons filled with timber. We used to call these little steam shunters "coffee pots".

The line to Lincoln also brings back memories. It was, like so many other lines, a victim of the "Beeching Axe" in the 1960s but I do remember that in the dying days of steam I used to travel on an early diesel railcar in dark green livery. This was a special treat because you could sit in the front carriage right behind the driver and see what he could see as he used his "deadman's handle" to control the power and speed. This handle was attached horizontally to a circular control and was so called because it would automatically shut off power if the driver collapsed. The route to Lincoln was littered with memorable station names like Stixwold and Bardney which, for some reason unrecalled to me now, lent its name to the expression "Do you come from Bardney?" – a phrase which would be directed at anyone leaving a door open in our house.

Train spotting fascinated me and I used to sneak off at the weekends to catch the train to Peterborough North Station and the main line

Driving the Flying Scotsman at Tyseley

between Kings Cross and Edinburgh. This used to be a regular route for the "Flying Scotsman" (which in later life I got to drive short distances at Tyseley near Birmingham). As my beloved steam fell into decline to be replaced by more reliable and cost effective diesel engines like the Deltic class named after race horses such as "Nimbus" and "Tulyar" I became less interested because diesel did not have the same romance and character as steam. In later years I realised of course that the economics, cost and reliability of engines that had to be prepared and maintained for a couple of hours every morning before a journey could start and required both a driver and a fireman to keep shovelling fuel into its firebox and the regular stops for water made diesel and electric trains an inevitable necessity.

 In my train spotting days until my mid teens I "copped" (the word for spotting the number and crossing it off in the Ian Allan train spotting bible) mostly engines from the LNER with the occasional scoop of a guest engine, so it was a real treat when my grandfather took me to Paddington station where I saw all the marvellous (and quite different)

GWR engines from the legendary "King" and "Castle" classes. In later years I was able to treat myself and my Uncle Len (who started his working life on the railways) to a "train driving" experience on the preserved Great Central line driving the former GWR engine "Clun Castle". The other excursion onto new train spotting territory was when I went on a camping trip with the boy scouts to Chapel-en-le-Frith in Derbyshire. On this occasion our whole scout troup went by train through stations such as Dore and Chinley, which sounded very special to my impressionable teenage mind. It was on this trip that, as was my usual practice, I stuck my head out of the window to look at the "Black Five" locomotive that was hauling us through what was, to anyone from Lincolnshire, mountainous Derbyshire countryside and got a spark in my eye so I spent the whole week with eyes watering. In those days there were separate compartments with "string" luggage racks and windows that used a thick leather strap to raise and lower them like a belt with metal eyes that slotted over a brass peg on the window frame. As soon as I got back to Boston my parents took me to hospital where a doctor used an instrument that looked uncomfortably like a potato peeler to extract the grit from my eye.

As I grew up through the 1950s and early 1960s and steam locomotives became consigned to the Barry Island scrapyard, Lord Beeching wielded his axe and many lines were closed, including the one that ran past our house. That track now carries the A16 road from Boston to Spalding and Market Deeping and the only line left unscathed in Boston is the route from Skegness to Sleaford and Grantham, primarily used to access main line services to London and the North via Grantham and to handle the seaside holiday traffic to Skegness in the summer.

It wasn't until many years after the last scheduled steam hauled passenger service made its sad progress along the iron road that I got the chance to actually drive a steam locomotive and came to understand something of the skill and passion that made being an engine driver the early career choice of small boys. I am now lucky enough to live close to Britain's longest twin track main line preserved railway, the Great Central which was actually the last main line to be constructed in the closing years of Queen Victoria's reign. This railway was, ironically, probably the most visionary and worthy railway construction project the UK has known, and a reason why some diehards still campaign for the Great Central track to be resurrected commercially. The Great Central was constructed not only to compete with the London Midland and Scottish (LMR) and the London and North Eastern Railway (LNER), in the days before nationalisation into British Railways, on speed and journey times on routes from Sheffield to London, but was also intended as a gateway to link to a future Channel Tunnel that was still a pipe dream.

The Great Central route made extensive use of bridges, embankments and cuttings to ensure that, unlike the Lincolnshire railway that ran past our house in Wyberton, there were no level crossings and therefore no need for manual labour to control crossing gates and vastly improved safety potential. In Leicester you can still see the old Great Central station and the many embankments and bridges that carried these magnificent machines through our cities and countryside. The flat that I live in now is on land that was used by the Great Central for sidings and engine sheds as the route passed within one hundred yards of my front door. Happily, a large section of the railway between Loughborough Central Station and Birstall (now called Leicester North) has been saved and there are plans to extend the route north to connect to the other half of its preserved section and Nottingham.

I had already enjoyed a number of nostalgic steam hauled trips from Loughborough to Leicester when I got to know about the "train driving" experience that was offered there. I paid for myself and my Uncle Len to share what is known as the "Bronze" experience consisting of some basic theory and introduction to the history of the Great Central before being let loose on real "live" locomotives. We arrived very early and with great excitement for a morning which began with a Railwayman's breakfast in the station buffet and a bit of driving practice on a shunting engine near the engine sheds at the rear of the station before we were taken on board the former Great Western Express locomotive, Clun Castle. Under the careful instruction of a now-retired former British Railway driver and with the firebox constantly replenished by a volunteer stoker, Len and I were able to share the experience of driving a passenger train along the route of this historic main line.

This experience really whetted my appetite and I jumped at the chance of two further experiences which were even more special. The first of these was a birthday present and gave me the dream chance to drive the actual legendary "Flying Scotsman" A3 pacific engine known throughout the world for its speed and beauty. This fantastic experience took place at Tyseley near Birmingham, once a massive centre for steam locomotion and now being used for preservation initiatives like this. Sadly, I could only drive the Pete Waterman owned pristine condition and original liveried giant just a hundred yards backwards and forwards along some sidings without any coaches attached and the time was also shared by driving a "Thomas the Tank Engine" shunter up and down the same distance. It was enough however to enable me to say with pride "I've driven the Flying Scotsman !". The power of this locomotive is such that you only needed to open the regulator (the handle which effectively lets steam into the pistons) a short distance and then almost immediately close it for the loco to have enough power to travel that one hundred yards.

    Neither of these two great "fantasy becomes reality" adventures could compare to what eventually came to pass when my beloved "Oliver Cromwell" came to the engine sheds in Loughborough to be rebuilt and restored to its former glory. Once I learned this, I kept an eye on the rebuilding progress through the local papers and the occasional visit to the engine sheds until they finally announced the chance to drive this locomotive. I wasted no time in booking my favourite loco for the whole morning, making the roundtrip from Loughborough to Leicester twice and sharing the driving with some of my closest friends and giving them an opportunity to travel in the carriages she was hauling. I had tears in my eyes and lump in my throat as all my boyhood memories came back as I finally had the chance to ease the regulator that made Oliver Cromwell smoothly pull away from the station with its wonderful hissing and powerful chuffing. This locomotive in its day was the ultimate in precision railway engineering with roller bearings that not only reduced maintenance costs but made the engine glide effortlessly along the track on its own momentum, so much so that on one trip back from Leicester North, my instructor driver told me I could shut the regulator down whilst we were still over a mile from our Loughborough Central destination as the slightly downhill gradient and momentum of locomotive and carriages would be sufficient to carry it that distance.

The Days of the Bomb and other Road Warriors

There were no cars in our street in the 1950s apart from "The Bomb", no garages or drives to park cars in and, only the weekly visits from the "Corona" lorry, ice cream van and occasionally, in the winter months, the mobile fish and chip van. The daily milk delivery came from a battery assisted milk float that the milkman used to pull down the street. The very first car I ever rode in, apart from The Bomb, was the taxi operated by the owner of the local corner shop. This Aladdin's cave of a converted front room was run by Horace and Nellie Pickett (even their names seem as from a bygone era). Nellie would look after the shop whilst Horace ran this taxi service that we used to take us to the station. The taxi was a large car with a gear shift on the steering column and comfortable leather upholstery. Horace would drive at a very leisurely speed smoking his pipe and listening to brass band music on his in-car cassette player. Looking back now, I suspect that this car was the love of his life and must have been a very early example of an in-car entertainment system. The fare for the two mile journey from our house to the station was two shillings and six pence, equivalent to 10 pence in the decimal currency that was introduced in 1971.

The other transport of the era was the local bus which ran the sixteen miles from Spalding to Boston along the old A16 and was operated by Lincolnshire Road Car whose bus depot on London Road it would pass on its way into Boston. I used to go past Nellie Pickett's shop on the way to the bus stop at the crossroads of Tytton Lane and London Road. Invariably I would be tempted to venture into Nellie's shop for a penny bag of broken crisps or, occasionally, sherbert flying saucers or a stick of liquorice. These buses were double deckers and I always hoped the my Uncle Len (who had left the railways to join Lincolnshire Road Car) would be the conductor as he would always let me sit on the "ice cream tub" which is what I called the space at the back meant for luggage.

Our scout troup also used bus transport to take us on camping trips, mostly to Woodhall Spa where the Boston and District Scouts had a permanent camp. These buses, operated by Cropley and Son, were very reminiscent of the ones used by Anthony Hopkins character in the film "84 Charing Cross Road". Our scout master seemed a very tall man –

especially to a little eleven year old scout, and it was therefore surprising that his car was the amazing bubble car manufactured by the same German company that made wartime planes, Messerschmidt. The canopy or bubble used to fold sideways to allow his lanky frame to slide in before he was cocooned and ready to set off. As I recall, the steering was effected by handlebars rather than a steering wheel.

It was my uncle's Humber Snipe "Bomb" that provided most of our transportation at weekends and on annual camping holidays to Hunstanton. It was started by using a crank handle, had rear doors that opened the opposite way to today's cars, and orange indicators that flipped up from the bodywork when the driver wanted to turn left or right. "The Bomb" was quite temperamental but it did manage to carry all 4 adults and 3 kids each year to Hunstanton and tow a little two wheeled trailer with all our camping gear. Uncle Bill had somehow "acquired" a military bell tent with a large central pole like a circus tent and all seven of us slept in a circle with our feet pointing to the centre. There were many happy days spent in Hunstanton's Searle's camp site only marred by a tragedy one year when some very rough weather and seas cost the life of a bather.

Our relationship with the transport technology called "The Bomb" was one of deep affection but cars in those days had no customisation options – "you could have any colour you like as long as it was black" was the Henry Ford motto of the day. Despite this lack of customisation, the less refined manufacturing techniques of those times and the need for constant care and maintenance meant that there was a real bond between driver and car. Drivers who loved their cars would care for them often better than their wives and partners and drivers would know by the sound of the engine ticking over or the vibrations when driving if there were any problems and often their cause and solution. This bond was something of a personal relationship between man and machine which enabled some drivers to coax extra performance and reliability out of their vehicles. This kind of relationship probably mirrors that between humans and domesticated animals where the right kind of subtle bond can bring special responses from the animals. Today you can see the same phenomenon articulated by Formula One racing car drivers though they now have a battery of telemetry and formidable technical team to support them.

My father only had a Raleigh "push bike" to take him to his work as a carpenter, joiner and undertaker employed by Addlesee and Son, the Boston funeral directors. Like so many of his generation whose youth had been snatched away by the Second World War, he wanted his son to have the opportunities in life denied to him and so he and my mum would make sacrifices to give me the best possible start in life. My mum worked at the local primary school as a school meals supervisor and, until I was old enough to have a bike when I started Grammar School, I used to either walk or catch the bus.

My shiny red Raleigh bike not only carried me to school but also took me at weekends in the summer, along with the extended family of my Uncle Bill, Aunt Iris, my cousins Mick and Steve and their neighbour's son Pete Brown, to Frampton Marsh with its creeks and mud flats and a lagoon we nicknamed the "Fairy Dell". The two wheeled cart which transported all our holiday gear to Hunstanton behind the Bomb was now towed behind my uncle's or my dad's bike and re-purposed to carry the bell tent, picnic gear and cricket set that would provide us with hours of fun and entertainment in the summer sun. It was here in those muddy creeks that I first learned to swim and on the sun-baked mud of the marsh that we used to play cricket. Frampton Marsh was adjacent to both the River Haven and River Welland and at high tide we used to watch ships apparently gliding by on dry land as their hulls and the waterline below were obscured by the tall river banks that were built to reclaim the marshes for rich farm land.

My Raleigh bicycle also influenced both my sporting and my love life. One of my pals at school was called Willie Worraker. He was cleverer and more academically minded than me and I regarded him as a bit of a swot. Willie's father was an art teacher at the Kitwood Boys Secondary Modern school and even as we competed at the top of our class we became good friends and he was probably my first best mate. His family invited me one year on their annual caravan holiday at Chapel St Leonards near Skegness. We were both around 14 at the time and beginning to take a romantic and somewhat naive interest in girls. On this caravan holiday, the first time I had stayed with another family, Willie and I met two young girls around our age, Denise Bowley of Appleby Magna and Heather Lees who lived in nearby Moira. With my winkle-picker shoes and Denise's "Kiss me Quick" cowboy hat we made a fine couple and those first innocent kisses sparked a year-long correspondence with love letters marked with SWALK (Sealed with a loving kiss) and BOLTOP (Better on lips than on paper) on the outside of the envelope. The growing strength of our promises to each other led to a commitment to meet again the following year.

Since there was no direct rail service to Chapel St Leonards and we didn't want to use bus or taxi, I persuaded Willie to cycle the 30 miles each way to see them for the day on their Easter holiday. This was by far the furthest that either of us had ever biked but the flat Lincolnshire landscape helped us to get to our gateway to a year-long much dreamed of paradise in under 3 hours. Once we had renewed our relationships with some chaste hugs and kisses we decided that a trip on the boating lake would be fun. My foolish inclination to show off to my much-missed sweetheart led to me stepping onto the rowing boat only for it to keel over and throw me unceremoniously into the cold April water. Since I had no change of clothing I finished lying in my underpants in the sand dunes whilst I unsuccessfully tried to dry my clothes in the chill wind by attaching them to a barbed wire fence. The cycle ride back to Boston in damp clothes was a trauma which not only signalled the end of a beautiful relationship with Denise but also triggered a bout of whooping cough. It was this whooping cough that scuppered my aspirations to play for England Schoolboys at soccer. I was selected for and did attend the England Schoolboys Under 15 coaching session at Lilleshall but I was quite ill and coughing up blood at the end of each session and so my talent was never spotted by the resident coach Dari O'Grady who managed Crewe Alexandra for many years.

The first motorised transport my dad ever acquired was a second hand moped we used to nickname "Phut phut" because of the noise it made when it sprang into life as a result of some fierce pedalling from my dad. Before I acquired my brand new red Raleigh bike with five gears, my dad had made me a go cart modelled in the style of 1920s racing cars. I called it the "Brooklands Flyer" after the famous pre-war racing car circuit and it is some coincidence that my mum has spent some years of her life in an independent living complex called Brooklands Gardens. My go cart was the envy of Yarborough road – it was controlled

My Lambretta Li150 – Passport to the Swinging Sixties

a steering rope attached to the ends of the front axle and powered by conscripted pals pushing from behind.

When I was sixteen my dad bought a Lambretta Li150 scooter – ostensibly for his own use but it wasn't long before I had laid claim to this wonderful off-white machine with sleek red panels. This scooter became my prized possession and was the gateway to a new social life and a girl friend to ride pillion with me. My Li150 took me to many new places and experiences during the fabulous "Swinging Sixties" when teenagers riding scooters like mine were referred to as "Mods" whilst the motor bike fraternity were "Rockers". Clashes between these two clans were frequent and immortalised in the film "Quadraphenia" based on the Who Rock Group album. I did not subscribe to being a "Mod" whose main uniform was the "Parka" jacket and my only concession to the "Mod" fraternity was a little lucky rabbits tail that dangled from the aerial at the back of the scooter.

Crash helmets were not compulsory in those days so I and my girlfriend Jenny would ride in the summer to Skegness with the wind blowing through our hair (in the days when I had hair). We were both into "Flower power" and I used to wear white flared jeans and a red silk Kaftan which I used to hide from my parents in case they thought I was turning strange. I did have one accident on my Li150 in the days when I used to play for Boston United Reserve team in the Lincolnshire League. To play for my home town alongside professional footballers had been a long held ambition and one Saturday morning I was riding down the one way street which ran past the Odeon cinema when the car on my left changed lane in front of me without warning. Fortunately there was a car park on my right hand side and I was able to avoid collision with this driver and swerve into the car park. Unfortunately there happened to be a bread van parked in my path and I went straight into its side leaving a "Tom and Jerry" like cartoon shape of a scooter in the side of the van. Apart from a few bruises I was unhurt and able to play football later that afternoon but I had to push the damaged Lambretta the 2 miles home and get my dad to sort out the mess. He managed to get it repaired at the garage he bought it from where the owner's son Derek Chatterton used to be well known competitor in the Isle of Man TT.

This was the age of the "Swinging Sixties", the Beatles, Rolling Stones, Pink Floyd and Carnaby Street. It was a fantastic era which saw the dawn of pop festivals like Woodstock, the Isle of Wight and the Bath pop festival (which later became Glastonbury. I was very fortunate not

The Boston Gilderdrome Dance Hall in 2011

MY NAME IS IVOR, I'M AN ENGINE DRIVER 27

Woburn Abbey – My first Pop Festival

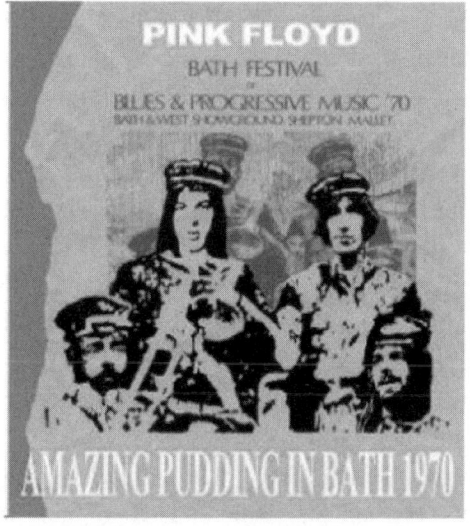

Bath Pop Festival where I saw Pink Floyd's Amazing Pudding which was later to become Atom Heart Mother

only to be living in a town which boasted the Gliderdrome dance hall and became the venue where I saw almost all of the big pop stars of the day. The Gliderdrome was the initiative of the local entrepreneurs the Malkinson Brothers who also owned Boston United Football Club that stood at the rear of the dance hall.

My Li150 was also able to take me to gigs around the country, including one of the very first pop festivals – held at Woburn Abbey.

Here I saw bands like The Move and Zoot Money and his Big Roll Band. I also went to a festival at Lincoln City Football Ground to see the Who with Pete Townsend at his guitar smashing best, Peter and Gordon and the Scaffold, who had a big hit with "Lily the Pink", but the finest pop festival would have to be the first Bath festival where I saw my Favourite band, Pink Floyd, perform in the early hours of the morning lying under the stars and hearing the ethereal sounds of "Astronomy Domine" and "Set the controls for the heart of the sun" swirl above my head. It was at this festival that they performed a piece of work called the Amazing Pudding accompanied by an orchestra. This later became their iconic "Atom Heart Mother" Album.

It was in 1967 that my life working with technology really began to take shape. I had long since abandoned the idea of becoming a train driver after the demise of steam and, since my parents were not well off, I sought to find an opportunity that would give me a funded scholarship to go to university. Engineering seemed to offer the greatest potential and I went on visits to the Rolls Royce Aero engine factory in Derby and remember seeing the wind tunnel where they were testing the newly invented carbon fibre turbine blades by hurling frozen chickens into the engines to see if they would be able to withstand a flock of birds in operational conditions. I also ironically visited what was the AEI (later Marconi) factory in Leicester, not far from where I now live. But it was Post Office Telecommunications (now BT) that gave me the chance that shaped my life.

Other members of my family had worked for the GPO (General Post Office) – my uncles Bill and Ray had worked on the telecommunications side and my uncle Craig became the Postmaster of the Oakham Sorting Office. I had my interview for a student apprenticeship scholarship in London near the Holborn tube station and during the train journey down I read a book I bought on the platform at Boston station. It was a guide to digital technology and it almost certainly helped me to get the job as the Post Office were switching from analogue to digital technologies and I got asked about the new System X during the interview. I remember celebrating the end of the interview by buying the latest Jimi Hendrix single "Hey Joe".

The relevance of this chapter of my life to transport is that it brought the biggest challenge to my Li150 (and me). As part of the one year pre-university induction training, all of our student apprentice (SA67) intake had to do a workshop practices course in Bootle, Liverpool, some 160 miles from Boston. This 6 week training in mechanical engineering began around October and I took myself to the home city of the Beatles on my Li150 wearing the warmest clothes I could find. I had no experience of finding somewhere to stay in a strange city but it didn't bother me that much as I headed to find a bed and breakfast. My first couple of nights were spent staying at a B&B in the legendary Penny Lane of the Beatles fame but this proved far too far from Bootle. Once on the course with my new new-found Post Office colleagues I discovered that one of my number, Mick Gulliford, lived in Ashbourne. My route from Boston to Liverpool had taken me along the A52 through Nottingham and Derby so we agreed that a practical solution would be for me to drive to Ashbourne and join him in his Morris Minor to continue the journey to Bootle where we managed to find a one-roomed bedsit in Merton Road, Bootle. This was a great arrangement both practically and financially because the Post Office paid good daily subsistence rates for such stays away from home. Our cheap bedsit meant that we were both in profit from this arrangement. This tiny room had 2 beds and a simple electric tabletop cooker with a small oven and two electric rings. Thus followed 5 weeks of a daily diet of Fray Bentos steak and kidney pie, "Smash" instant mashed potatoes, baked beans and coffee which was a perfect combination for the little cooker and meant that we could chat up the counter girl in the local Woolworths store where we bought this savoury fare. It was whilst I was in Liverpool that I saw one of the incredible road shows that used to tour the country playing in cinemas and concert halls. Each band or artist would only play one or two numbers before the stage revolved to bring on the next act. Seeing Pink Floyd and Hendrix and The Nice in one evening is the stuff dreams are made of.

Even when I was wearing two jumpers and a huge heavy woollen overcoat I used to finish each weekly journey (we both went home at the weekends – I was playing for Boston United Reserves on a Saturday and Gipsy Bridge on a Sunday morning) totally shivering from head to foot. It was during this one year period before University that I was also

The PInk Floyd/Hendrix roadshow

introduced to another contemporary form of transport – the motorbike. One of my other fellow student apprentices was a character called John Backhouse, a highly intelligent Lancastrian with a build like Herman Munster and a wide forehead to match. I often imagined that he had bolts on his neck hidden underneath his shirt. It was during one of our other induction courses that a group of us, including John, chose to stay at the YMCA on Tottenham Court Road, some miles from our training course at Dollis Hill, but ideal from a financial point of view giving maximum profit from our London weighted subsistence allowance. John owned a much loved motor bike and I was the only one on the course bold enough to agree to be his pillion passenger on the journey along Edgware Road in the morning rush hour.

John love to drive to the limit, overtaking all the cars with the bike's foot pedals occasionally showering sparks as they scraped along the concrete kerb of the central reservation and me clinging on in fear of my life. We did have one accident however when a lady stepped out in front of his speeding bike on a pedestrian crossing. He happened to be on the inside lane at the time and the zebra crossing was one of the now largely defunct belisha beacon types (*according to Wikipedia "A* **Belisha beacon** *(bəˈliːʃə/)) is an amber-coloured globe lamp atop a tall black and white* pole, marking pedestrian crossings of roads in the United Kingdom, Ireland and in other countries (e.g., Hong Kong, Malta) historically influenced by Britain. It was named after Leslie Hore-Belisha (1895–1957), the Minister of Transport who in 1934 added beacons to pedestrian crossings, marked by large metal studs in the road surface. These crossings were later painted in black and white stripes, thus are known as zebra crossings. Legally pedestrians have the right of way (over wheeled traffic) on such crossings.) I used this black and white pole beneath the flashing amber globe to swing off the bike in "Tom and Jerry" style just before his bike skidded sideways to the ground. No-one was injured in this incident but John was beside himself at the scratches on his paintwork.

John Backhouse was something of an inventor and one of his gadgets was a converted electric blanket inside his leather jacket which he rigged up to be powered by electricity from his bike. Fortunately the Dollis Hill course was mercifully short, unlike some of the training we had at an isolated country mansion called "Horwood House" near the village of Little Horwood and not too far from Bletchley, home of the famous Bletchley Park that played such a key role in the wartime by hosting the special team of mathematicians that broke the German "Enigma Code". This was in the days before the city of Milton Keynes existed and Little Horwood was one of the hamlets to be eventually swallowed up in this massive development. The other memorable journey on the back of John's bike was when I got permission to have an afternoon off to play in a mid-week floodlit game at Boston. John took me to Bletchley station to catch the train home and it was one of the scariest trips I have ever known with John driving at what seemed to me breakneck speed down these winding rural lanes, leaning hard into every corner with, on occasions, sparks from the footrests as his bike banked left and right.

Boats

My first maritime transport experience was bizarrely on board a Russian vessel in Boston Docks where I sampled caviar and vodka for the first time. I think it most probably was a timber ship from the Baltic and the reason I was on board with some chums from Grammar School was that we had Russian language lessons taught by "Gert" Tromans who later also organised an amazing school trip to Moscow and Leningrad (now St Petersberg). Rather strangely "Gert" also taught religious education to "O" level standard and this was the only subject I failed in these exams.

The school trip to Moscow and Leningrad in 1966 was also my very first experience of being abroad. My parents saved hard for me to be able to afford to go thinking it might be rare lifetime chance to experience countries outside the UK. Little did they or I have any inkling how often I would travel later in my life or how easy and cheap it is now to travel abroad. The seventeen day school trip joined by about a dozen pupils and led by "Gert" Tromans began with a rail journey to London and tube to Victoria Station where we caught the "Boat Train" across the Channel. It was at Victoria station that I discovered a "never-seen-again" recording booth that actually made a 45rpm vinyl record of any piece you cared to record. Always up for a challenge, I sang something and took the recording to Moscow where I managed to sell it for a rouble along with other rare capitalist items like "Bic" ball point pens. The journey across Europe was by train and amongst the highlights were the crossing of the Berlin Wall by steam train when armed guards boarded the train to confiscate any food we were carrying, crossing the noticeably poorer regions of East German and Poland to the Russian border at Brest and the final leg into Moscow.

It was in Moscow that we stayed at a hotel which overlooked Red Square. I believe that this hotel must have been the Rossiya which, on my last visit to Moscow in June 2011 was still being developed. Russia seemed very poor at that time with most shops having empty shelves and lots of beggars on the street. Whilst in Moscow our party, as was the practice for all foreigners, was taken straight to the front of the queue for Lenin's Mausoleum and we all filed past his wax-like body in its coffin. Our trip also coincided with the 1966 World Cup which we watched in ghostly, flickering black and white images on the one television in the hotel. The Russian commentary and the locals watching alongside us left us in no doubt that the Russians wanted England to win, even if only because of their lingering post-war hatred of the Germans.

From Moscow we took the train to Leningrad where amongst the highlights was my first (and only) trip on a hydrofoil boat along the river and our visit to St Peter the Great's spectacular Summer Palace. The trip back to the UK was concluded by the cruise ship SS Baltica to Tilbury Docks via Helsinki and Copenhagen. We visited the Tivoli Gardens in Helsinki and saw the "Little Mermaid" in Copenhagen before the final leg home to London and train back to Boston. This amazing voyage seems much of a blur today with half-formed images in my mind but I do remember one phrase used by one of my school mates to insult a fellow pupil for an act of stupidity (and there were many on this trip). "Conical-headed apple-cheeked twit" is the expression that was so appropriate at the time that I am still waiting for another act of stupidity from someone who resembles that description to be able to reprise that phrase as my own !!.

If my Raleigh bike and my Lambretta scooter were passports to romantic experiences, various motor cars should not also go unmentioned in this context. I had no car of my own until I left university but during that one year pre-university industrial training I befriended at least 3 car owners that brought such dividends to my love life. The first car owner was a Geordie called Colin from Sunderland. When I met him at the Telephone Exchange in Boston he was a contractor installing new switching equipment. Colin had, at that time, a Chrysler Hillman GTi. He was a real charmer and after he had impressed my parents with his special way of making coffee by blending the coffee granules with milk and sugar into a paste before pouring the hot water, they had no hesitation in letting me go out on the town with him. My abiding memory of one such outing was a trip to an RAF dance at Coningsby where we managed to persuade 2 girls to give them a lift home with Colin in the front and me in the back. I remember kissing my all too willing partner to the sound of Fleetwood Mac's "Albatross" as Colin wooed the other lady with his in-car stereo cassette player.

Colin also introduced me to the Star Inn at Sibsey where he moonlighted in the evenings behind the bar. This pub had been renovated to be modelled on the style of a wild-west saloon complete with hitching rail. It was the brainchild of a Nottingham entrepreneur called John Lord and was quite an unlikely commercial success because it was miles from the nearest village and only benefitted from visitors travelling along the A16 from Boston to Grimsby. John Lord organised sell-out cabaret evenings including one strip show which made the inside pages of the News of the World because it attracted so many customers that most of them were forced to stand outside peering through the large plate glass windows. Colin also managed to get me a part-time job behind the bar at the Star Inn where I met another vehicle owner whom I rescued from being manhandled by a drunk in the pub doorway. I threw the considerably shorter-than-me man out and became friendly with his victim, a girl called Wendy who owned an Anglia Estate which she used on one occasion to drive us up to the Yorkshire Moors for a long romantic weekend where we slept in the back of her car in separate sleeping bags on a bluff overlooking the North Yorkshire preserved Railway line near Goathland, home of the much later and much loved TV series Heartbeat.

My relationship with transport technology and the importance of it in my life continued once I started my Electronic and Electrical Engineering Degree course at the University of Birmingham along with just two of my fellow student apprentices, "Jim" Ames and Don Fisher. At this time, very few students owned or had the use of a car and Don was one of the lucky few and it earned him a passport to social popularity. By this time I was back to my bicycle although most of the time I walked around the campus. Don became a friend once we started this course and shared my interest in rock and dance music. Don became my "roadie" in my final year when I started my own mobile disco and launched the "Progressive Underground and Rock Society". With the help of Don and his car I managed to get the legendary radio DJ John Peel to become honorary President of this fledgling society which held regular meetings in the student union. John Peel was doing a gig at Brixworth in Northamptonshire alongside the band "Earth Wind and Fire" so Don and I set off from Birmingham to meet him. He was extraordinarily gracious in accepting this request and I still possess a postcard from him asking how the society was going.

The society really took off when we ventured to organise a film night and hired a room, 35mm projector and two rare films featuring Jimi Hendrix and Pink Floyd. I had some posters printed offering tickets at 2 shillings for members and 4 shillings for non-members. I had some posters printed in my home town and I needed about 80 people to break even. The room we booked held 100 people and we had organised a second room for me to run my disco after the films had finished. Incredibly we sold over 600 tickets and had to show the films 4 times to accommodate everyone. It was a great financial success that funded most of my record collection from that era. My venture into the mobile disco business was assisted by my dad using his woodworking skills to build me a cabinet which housed 2 record decks and could carry amplifiers and speakers as well as my record collection. For most of my gigs at the university I had no need of additional transport technology because the cabinet was on castors so I used to wheel it out of my room on the 17th floor of High Hall student residence, into the lift and along the campus to wherever the booking was. Most of the bookings were at my own Hall in basement room LG4 where my Sunday night discos were regular sell-outs, increasing my popularity rating with everyone except my tutors who felt that running a mobile disco and playing for the University football team were incompatible with getting a good degree.

On the occasions when we were booked for private parties, Don would hire a minibus that we would cram all the gear into and it was the minibuses that played another pivotal role in my sports life as they were regularly used by the University soccer team. I became the treasurer of the University Soccer club and looked after the finances when we once went on a summer soccer trip to Heidelberg. One of the second team player's mother had made us all special red jackets with BUAFC in bold white letters on the back and the whole team travelled to Germany in a minibus driven by our goalkeeping captain Ian "Ted" Finch. This responsibility for the club finances on the trip helped me to learn to count in German and to order in German from the menu. Most of our meals seem to comprise of "Ein Haupt Henchen mit frites" or half a roast chicken and chips.

It was whilst I was at university that I also had my first experience of air travel. In my final year I had met a girl called Pauline "Polly" Feather who was studying combined French and German at university. When Polly started her year abroad with some months at Lyon University and an equal period at Freiburg, I suddenly decided it might be a nice romantic gesture to pay her a surprise visit one weekend. In my rashness, after booking a flight from somewhere in Kent to Paris and the rail ticket from Paris to Lyon, I had almost no money left when I struck out from Birmingham. Not only did I have little money, I did not know her address in Lyon. I arrived in Paris in the early evening and somehow managed to navigate the metro to get to Gare de Lyon to catch the night train to Lyon. For the whole journey I dare not fall asleep in case I missed the stop and arrived in Lyon around 4 am in the morning. I slept on the station bench until it was light and then consulted the map in the station booking hall to see where the university campus was and then set out walking there. Incredible as it seems now, especially with the number of students in Lyon, I found her almost immediately as I queued for breakfast at the refectory. When she started her course in Freiburg I paid her another surprise visit but this time I knew her address, etched still in my memory today – 3 Rasenweg !!

Transport technology came into play again when, before our final year at University, we could choose a foreign country for an exchange visit to study their telecoms services. I was lucky enough to get my first choice of Norway accepted and so I travelled to along with a colleague to Oslo to meet my host and we were driven as far North as Alesund at the edge of the Arctic circle before travelling down the west coast along the magnificent fjords. We mostly stayed in youth hostels during the two week trip and the most memorable incidents were eating reindeer steak and cloudberries and cream in a mountain refuge, sailing down a fjord and visiting the Olympic ski jump site at Lillehammer before we set sail from Bergen for home.

It wasn't until I left university and started work at the GMO (General Manager's Office) headquarters in Leicester that I was able to afford my own car – a second hand Triumph Herald Estate chosen mainly because it was able to carry my disco gear in the back on the occasions when I ran discos for my friends. I took out a loan of £600 to buy "Dulcie" – so-called not because I thought of her as female but because her registration was DUL 814C. It was with Dulcie that my first really meaningful relationship with transport technology started. Whether it was because she was the first machine I had actually owned or because I used her regularly for pleasure trips that I felt this bond I couldn't actually say. I am certainly not a "petrol-head" or a man that loves to spend hours polishing and servicing their car – I have always seen cars as something to get you from A to B. However, there was definitely a relationship with Dulcie which illustrates to me how machines and technologies have evolved to where we are today. When I started up and drove Dulcie I seemed to know instinctively how she was feeling and how she would perform. When she was manufactured it was a time when the human dominated factory processes and techniques of the day (pre-robot) meant that every car was unintentionally slightly different and all cars required careful "Running-in" for about 500 miles with regular services needed every 3,000 miles.

This individuality and need for careful handling meant that cars actually responded better when treated with care, skill, sensitivity and understanding and that even drivers like myself could sense when they needed restraint and when you could drive flat-out. I do remember one occasion when I was driving up the A50 hill towards Markfield near Leicester with a full load of disco gear on board that I felt Dulcie really struggling. I handled her gently up this long but not too steep incline and I remember thinking to myself "Come on old girl, you can make it". The point about all this in relation to the theme of this book is that my wooden-dashboarded Dulcie, like all cars of her time was an un-personalised, "One size fits all", unintelligent machine where nothing about the car had any intelligence to understand or learn about my needs. My relationship with Dulcie was a one-sided almost unconditional love affair in which I modified my natural behaviour to try to get the best from her. Very few things could be more illustrative of the changes in transport technology than the developments in in-car intelligence which are embedded in my current Fiat 500 that I will refer to later in this chapter.

It was the year after I started work in Leicester that I got married to Angela, whom I had met at Birmingham. She had spent a year studying on a post-graduate teacher training course at Loughborough Colleges and I was playing part-time professional football for the now defunct Loughborough United in the Midland League. When we met again in Leicester I was sharing a bed-sit in Gimson Road Leicester before we moved out to a rented flat in a farmhouse at Dingley near Market Harborough. This was to be closer to Angela's job as a P.E. teacher and the Robert Smyth Grammar School where the former Tigers and England rugby captain Martin Johnson was educated. From Dingley we moved back to another rented ground floor flat in Stoneygate Road, Leicester before we eventually bought our first house, a brand new three bedroom "Fletcher Homes" detached property in Cromwell Road, Great Glen for the princely sum of £10,000. I learnt later in life that one of our neighbours was Englebert Humerdinck who lived in the mansion whose grounds were just the other side of the wall that stood opposite our house.

It was during the first year of our marriage, which took place at Barnsley Town Hall just after Xmas 1972, that Dulcie was called upon to help a family crisis. Soon after we were married, Angela's mum had been diagnosed with cancer and her father, who worked in the Sheffield steel factories had a heart attack brought on by the stress of trying to care for her. He died shortly afterwards, which postponed any potentially life-saving treatment on her mum, and so it was that for almost a full year that Dulcie would carry us up the M1 to Barnsley every weekend to see her mum and family before eventually and sadly she died. Those weekly journeys were very sombre affairs and it was sad to see a loved-one's life slip away so slowly and painfully.

It was Angela's first job at Robert Smyth that introduced me simultaneously to package holidays and skiing. It seems hard to believe now that I was less than enthusiastic when she suggested I might like to join her school skiing trip. In my defence, I think that my lack of enthusiasm was more to do with being on holiday with a crowd of adolescent school-kids than an aversion to sticking two planks of wood on my feet and risking life and limb at the top of some distant mountain. The resort chosen for this holiday was Mitsurina in the Italian Dolomites. It was a quaint village with spectacular views of the mountain scenery that I so came to love over the ensuing years. I remember little of that holiday except that that it was an introduction to package holiday flights and the start of my lifelong romance with ski holidays. I had no suitable ski clothes and recall being in jeans and a fleece top for at least my first two ski holidays.

In those days, many of the flights were operated by Dan Air (or Dan Dare as we used to call them). We flew from Luton Airport which became a regular starting point for future such holidays. It was the package ski holiday that was probably responsible for my enduring love of air travel.

Even the interminable queues at the check-in desks with crowds of people, luggage, skis and fretful children became an almost welcome appetiser to the moment when you are finally on board and the jet soars above the cotton wool clouds and you hear the soft "shhhhhhh.." of the engines and your body and mind relax as you anticipate the pleasures to come at the end of your journey. It was probably also the infrequency of travel abroad, limited as it was to the annual trip to the mountains, that added to the sense of excitement and anticipation. The following two years after this initial trip we went to Solden and the fabulous Austrian resort of St Anton. It was our good fortune that the hotel that had originally been allocated to this package was over-subscribed and our initial fears when we were the last two people on the bus as it made it's way through this magical alpine village turned to unrestrained joy when we found ourselves in the best hotel in St Anton, including my first experience of "intelligent" urinals that used infra-red to detect when you had moved away and automatically flushed the basin. This led to us using the expression "I'm just going to break the beam" as a euphemism for "spending a penny".

St Anton's fabulous ski terrain was wasted on a novice like myself. There were no special short skis for beginners and our ski instructor Hubert (whom we nicknamed Heudebert after the continental crisp bread served at breakfast) did not let us use any ski lifts until the last day of our lessons. This meant that, still in my jeans and fleece, I had to step sideways up the nursery slope in my skis until I reached sufficient height to make a slow and hesitant snow plough descent. The last day's lessons bought the special treat of being allowed to use a button lift to drag us up the slope giving a much longer descent. In later years this annual one week holiday became two weeks, then four weeks and eventually to two one week holidays, one of which was a bachelor trip with my best mate Dave Timson. In those days, our social circle at the squash club was known as "Rentacrowd" as we were collectively guaranteed to liven up any party. There were many memorable Rentacrowd ski trips over the 1970s with incidents to provide enough material for a best-seller novel.

My love of air travel was rather unkindly attributed by one of my friends to an erotic sequence in one of the "Emmanuelle" movies of that era in which she and a fellow stranger passenger enjoy themselves under a blanket on a long-haul night flight, all to the aforementioned soft "sh....." of the jet engines. I hadn't been working in Leicester very long when one of my staff – a young man who had got his pilot's licence, offered to take me up for a spin – no doubt hoping it might improve his appraisal rating and promotion prospects. Since he had offered to fly me anywhere of my choice within the range of his Cessna, I asked if we could fly to my home town of Boston, some 60 miles away from Leicester. We took off from Leicester's Stoughton Airport, home of the local flying club, and I was allowed to help with the navigation and used my familiar railway lines to track us the last few miles until we got sight of the Boston Stump parish church and the docks. This trip meant a lot to me as my dad had been in the RAF so we had arranged for my mum and dad to stand in the back garden waving at us whilst my pilot waggled his wings in salute. It was an emotional moment amongst many others I have experienced whilst on a plane.

It was this early experience of being on a small light aircraft that encouraged me to take up flying lessons. The flying club at Stoughton was only a couple of miles from my home in Great Glen and the sight of small aircraft doing "the circuit" became a regular event as we sat in our back garden. My flying instructor was an ex-RAF pilot called Ian Sixsmith and, after I had passed my initial exams and spent enough hours in the sky, it was time for me to fly solo. It was an emotional and very scary moment for me when, after a few practice circuits with my instructor, I took to the sky alone. Strangely enough it was not the flying that I was apprehensive about, it was the dialogue with traffic control with its own special coded language. Once I had flown solo a couple of times, I invited my parents over for the weekend, not telling them that I had been having lessons and had already flown solo. It was sunny summer Sunday that weekend when I suggested that it might be fun for them to watch me have a pleasure flight, so they were left in the control tower clubhouse whilst I and my instructor walked over to our two-seater Cessna for the pre-flight checks. I explained to Ian Sixsmith that my dad had been in the RAF and didn't know I had been having lessons so once we had taken off and done a couple of practice circuits with "touch and go" practice landings, it was time for me to go off on my own again.

After we taxied to the control tower to let my instructor get out, I turned the plane round and set off for the main runway with Ian shouting up to my parents watching from the control tower balcony "I'm not going up with him again – he's a bl%^dy nutcase". My mum was beside herself with worry but they were both so proud and emotional when I did another couple of successful circuits before joining my parents in the clubhouse. That little practical joke led to me inviting my Uncle Len to the flying club on one of his visits. Len was so terrified of flying that he had to make himself intoxicated before he would even get on a plane and his only trips had been to visit my Aunt Jean's relative in Canada. I told Len that I was having a flying lesson and asked him if he would mind taking pictures of me and my instructor beside the plane. I had already tipped the instructor off about what was to follow. Once at the plane I said "It's not me who is having the lesson Len, it's you" and gave him no time to argue before he was strapped in and off on a pleasure flight which included him having control of the plane as it cruised over the Leicestershire Countryside. This ruse certainly paid dividends and did wonders for Len's fear of flying. Being at the controls of this relatively flimsy light aircraft made him feel much more in charge of his destiny when compared to his feeling of helplessness in the seat of a modern airliner.

 I could not afford enough lessons to extend my flying career to get a pilot's licence but the Stoughton airport and planes continued to play a part in my life through the annual air show which brought many different planes from all over the world to perform fly-pasts and displays. This included both the Red Arrows and the Battle of Britain memorial flight which is based at the Coningsby air base near Boston. My parents would both come over for this air show every year and we would all get choked at the sight of the Lancaster bomber flanked by the Spitfire and Hurricane as they flew low over the airfield. Other spectacular and emotional displays came from the Harrier Jump jets as they hovered like helicopters, Concorde as it did a mock landing, and the amazing Vulcan bomber whose sound as it opened up its engines to fly almost vertically into the sky had to be heard to be believed. It was at one of these air shows, on a particularly unseasonably chilly afternoon, that my father collapsed and had to be taken by ambulance with blue lights flashing with a suspected heart attack. Thankfully dad soon recovered the same afternoon and was able to return home with us in the evening.

It was my dad's love of aircraft that prompted me to book seats on a special Concorde leisure flight from East Midlands Airport. By this time I was divorced and living alone in a two roomed cottage very close to the business I had started on London Road, Oadby on the outskirts of Leicester. At the time I was dating a girl called Jenny whom I had met at the local church in Wigston and all four of us went on this trip from East Midlands airport to Heathrow with the return trip by coach. Once again I wanted this to be a big surprise for my mum and especially my dad whose last flying experience had been in the RAF on board a De Havilland Mosquito plane. I told them that Concorde was visiting East Midlands and that I had arranged a ticket to experience a walk round the aircraft. They were not even suspicious when we all had to go through a full security check before boarding the plane. It was only when we were on board that I revealed that she was actually going in the air and what an experience it was with champagne served and nice souvenir mementos to complete a special day which took in a supersonic section over the Atlantic and a simulated "touch-down and go" landing at the annual Fairford airshow. This special type of trip was repeated about a year later when British Midlands airways offered a leisure flight to celebrate the purchase of their new Boeing 767 fleet. Once again my parents were my guests as we flew over towards the Wash, including flying over Boston where we could even make out my mum and dad's house from the passenger window.

Air travel has not only brought enormous pleasure and rich new experiences as Thomas Cook envisaged in his ambitions, but it has also brought its share of scary moments, the first of which was on board a "Dan Dare" package flight for a ski holiday with my Rentacrowd friends. We had only been in the air for a relatively short time when the pilot announced he would have to return to the UK to make an emergency landing because of icing problems. He flew over the channel to dump aviation fuel and then went back to our starting airfield. Fortunately all turned out well and we landed without incident and only had minor delays once we took off again.

The second frightening experience was when a group of the lads from the squash club were persuaded to go on a charity parachute jump at Langar airfield near Nottingham. This involved a weekend course in which the Saturday was spent on classroom training that mostly consisted of preparing us for all the things which could go wrong on the jump. This focus on things going wrong did little to inspire either our confidence or our courage and included advice on what to do if your static line does not release your parachute and leaves you dangling behind the plane in its slipstream. The first advice is not to panic, a pretty difficult thing to do in these circumstances, and the instructor then went on to say "If you are conscious!!!!!! (alarm bells ringing), signal to your jump instructor stood in the open hatch of the plane and when he shows you his knife give him the thumbs up to signal that you are ready for him to cut through the static line." (diarrhoea fast approaching) after which you should remain calm enough to open the reserve chute strapped to your chest." It then got worse as he went on "If you are unconscious (I want my mum) the instructor will slide down the static line and sit on your shoulders whilst he cuts through the line and as he falls off your back he will release your reserve chute and you will have a fighting chance of cheating death." These guidelines delivered in the most matter of fact way did little to give a good night's sleep afterwards and were accompanied by describing how the plane would stall and plummet to the ground killing everyone if you were foolish enough to open your reserve chute whilst dangling at the rear of the plane.

The next morning we made our way to Langar airfield prepared to meet our fate. There was an unusual silence in the car and when we arrived we were given some exercises in falling and tumbling in various directions to prepare us for hitting the ground with our knees bent and rolling to the floor. As we practised these techniques as if I lives depended on it, our aircraft landed and picked up two women who had been trained the previous week but had been unable to jump because of poor weather. The plane circled upwards into the Nottinghamshire sky and we became transfixed as we witnessed a rehearsal of what we were about to do. Once the plane had reached its optimum altitude, between 1,000 and 2,000 feet above us, it began its run across the airfield. We saw one of the girls stood in the doorway ready for her jump as the plane's engines were cut, bringing an eerie silence. In this type of plane you had to step out of the doorway onto a step and hold on to the wing struts until the instructor tapped you on the shoulder and you released your hold to go into a star shape before the parachute opened. The first girl jumped and after a short time her parachute opened and she drifted across the airfield to land with rigid legs on the tarmac runway, falling back and cracking her head badly. The second girl seemed to lose her nerve and clung onto the wing strut only for the pilot to flip the plane so she was thrown off, eventually landing safely in a nearby field. These sights were far from confidence boosting as we checked each other's parachutes before it was our turn.

Four of us sat silently on the floor of the plane as it made its way higher and higher before we heard the ominous silence of the engines being cut. I stood in the doorway, grabbed the wing strut and positioned myself on the step. Now I had mentally myself prepared for the routine of jumping backwards into a star shape and shouting "1000, 2000, 3000, check" before glancing up to see if the parachute had opened. Despite all this repetition over the previous night's sleepless hours, as soon as I felt the tap on my shoulder I closed my eyes and jumped with a cry of "aaaargh". Suddenly I felt the pull on my shoulders as the static line opened my parachute and I sobbed with relief at the sight of its billowing canopy and actually enjoyed the glide over the patchwork scenery so familiar in our green and pleasant land. Even though we collectively vowed never to do it again, four of us went back a second time with much cursing as we drove to Langar but fortunately happy landings for us all. For some time after that second jump I would break into a sweat if I was sat in our Great Glen Garden and happened to spot a Cessna doing its circuits around Stoughton flying club.

These scary experiences did not deter me from looking forward to a Microlight flight given as a birthday gift some years later. The so-called pleasure flight took place at an airfield near Lichfield in a contraption that reminded me of a flying bathtub. There was no question of flying one of these solo so I went up with an instructor and once into the sky enjoyed the open air sensation of microlight flight, controlling the machine with a bar that was attached to the wings. This flight only lasted about an hour but was great fun.

It was some years after this that my then brother-in-law Dennis Pelling invited me for a spin in his plane which he kept at a small private flying club on top of a hill not far from where he lived in Beverly in South Yorkshire. Dennis was a very successful business man who lived with my second wife's sister in Bar House in Beverly – a listed building that was once home to the Monarch's artist in residence. Dennis was making money from worldwide rights (apart from North America) to the "Chips Away" car body repair franchise. He knew that I loved flying and had already had lessons, one crisp Boxing Day morning he suggested taking his plane up for a mystery tour around local flying clubs in the area. His four-seater plane was kept in a bubble hangar and we soon had it prepared for our flight. Dennis allowed me to take the plane up and as we climbed into the cloudless blue sky he started pumping a handle between our seats. "It's nothing to worry about" he said "One of our wheels has not come up so I'm pumping it down manually to try again – it's probably just the cold weather". Once the wheel was down again he pressed the button to bring both wheels up with the same result – one wheel remained resolutely down. He pumped again and informed me that he couldn't be sure that both wheels were locked so we had to fly to the nearest commercial airport where they had a tarmac runway and provision for emergency landings. This happened to be Humberside airport, not far away.

I continued flying the plane using his satnav system whilst Dennis radioed to Humberside to explain the problem and ask for fire crews to be put at the ready. He remained admirably calm throughout, which is just as well because it helped me to remain relatively unworried. As we flew towards the Humber bridge, black rain clouds started to appear and Dennis took control to bring us into land. The traffic control had directed us to land on the runway at an angle to the main commercial one so that if we crashed we would not hold up their regular and charter flights. Coming in to land, Dennis told me he would cut the ignition just above the runway so that if the wheels collapsed, it would only damage the propeller and not the engine (being a canny Scot he was trying to avoid unnecessary cost. I could see the fire engines lined up with lights flashing and the jet charter flights queued up on the other runway as Dennis suddenly pulled the cushion from under me and told me to hold it in front of my face in case we pitched forward if the wheels collapsed. It was some considerable relief to us that the wheels held but some disappointment to the fire crews as they followed us to a parking bay, slightly miffed that we had robbed them of some practice. Dennis offered to let me get a taxi back to Beverly but suggested that now we knew the wheels were locked we could resume our plans and leave the wheels permanently down, which is what we did and had a great day which included flying alongside a steam train as it made its way along the North Yorkshire moors track to Pickering.

Air transport technology from a passenger perspective changed little for those package holiday trips and it wasn't until my first long haul flight to Canada that I saw some developments which reflected the increasing use of technology to support personalised experiences for passengers. I had a won a study tour to Canada as a result of an internal competition within IBM and I and a colleague from the South, Clive Jecks, spent a week taking in Montreal, Toronto and Vancouver. Although the trip was a business study tour based around the IBM System 38 computer, we did manage to see the Montreal World Expo, Niagara falls and some of the countryside around Whistler Mountain north of Vancouver. I fell in love with Vancouver as we crossed the bridge at sunset and saw an amazing vista of city, sea and snow capped mountains. This trip also introduced me to in-flight entertainment systems which, at the time, consisted of drop down screens and a map display showing the journey progress.

My first introduction to more personalised in-flight entertainment with the now-familiar screens mounted into the seat in front of you was on a surprise vacation I organised for my second wife to be and her two children. I managed to organise a passport for her without arousing suspicion and took Carol, Stephen and Christopher down to Heathrow where it wasn't until they were at the departure gate that they realised we were all bound for Florida. I recall that we flew by Virgin Atlantic to Miami and had arranged for a car hire at the airport to give us the freedom to tour around Florida. Nothing was pre-booked but I was working on some guidance given by one of my staff, Rachel Smith, who had done the same thing some weeks before. It was the first time I had ever hired a car or driven in America and although the Corvette was a bit on the cramped side, we set off on a journey which began by heading down the Florida Keys. During those two weeks we went from Key West to Cape Kennedy, stopping on the way at a small town called Delray Beach, which later became the site of an unplanned timeshare acquisition on our return journey, and on to Orlando to enjoy the pleasure of Disney and

Epcot, and even over to the west coast of Florida for a short stay before driving back to Miami for the flight home.

The early in-flight entertainment offered a choice of movies but all the movies started at the same time and had no way of pausing or rewinding. This meant that to see the whole film you had to select it before the start time and remain in your seat until it finished. These systems also included the option to watch your journey progress across the Atlantic. Now married to Carol in 1989, we organised a second trip for the whole family of her four children and their partners to Florida to spend one week in our newly-acquired timeshare in Delray Beach and one week in Orlando. This trip was largely organised by my stepdaughter Nicki who booked our flights with a company called Pegasus, operating out of Stansted. Not only did the Boeing aircraft used on this journey not have any in-flight entertainment, it also only had two engines for an Atlantic crossing which I had thought was not allowed on safety grounds. We had to stop for refuelling at Bangor Maine before flying on to Miami.

The tickets for this flight contained the usual warning to call the airline 24 hours before departure on the return journey in case the allocated times had been changed. This trip was the only time I can remember when I actually did heed this warning and rang the day before we were due to set off. I was told that the departure times were still the same. You might imagine our feelings when we arrived at Miami airport over two hours before the scheduled time to find the check-in desk closed and our flight already taken off. There were eight of us stranded with all our luggage but I have to pay credit to Pegasus for the way they resolved this crisis. After we called the number on the check-in desk, an agent arrived and she managed to find an extra Virgin Atlantic flight to London Gatwick that had been necessitated through a fault on one of their aircraft. She booked and paid for all our seats and arranged for a minibus to pick us up from Gatwick and take us to Stansted. All's well that end's well as Shakespeare would say and we not only had a better flight with more leg room and good in-flight entertainment and food but we actually got back to Stansted only two hours after our Pegasus plane had landed.

It wasn't until 2006 that my regular international travel for my job started. It began with a long-haul trip to New Zealand for the ISDE

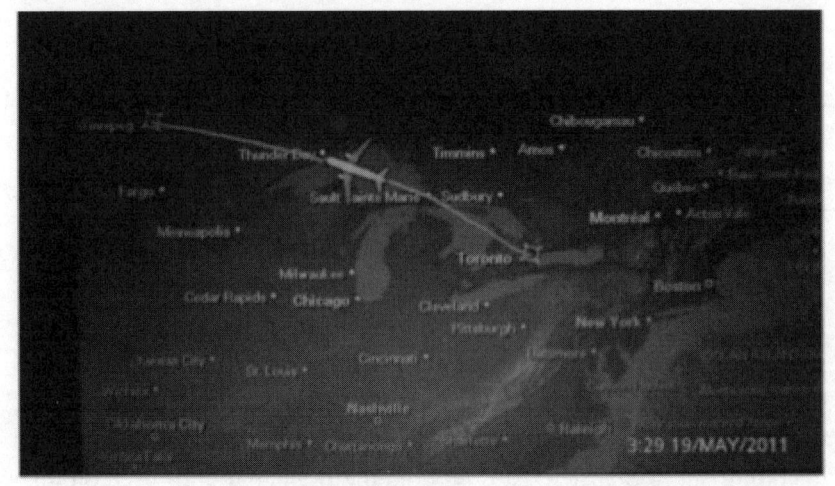

Modern Air Canada in-flight information system

Digital Earth conference in Auckland. I was working as a Project Manager for De Montfort University at the time and it was ironic that whilst I was in Auckland I responded to a job advertisement for the new Director post at the Serious Games Institute (SGI). Once back in England, I went for the interview, was successful, and in November 2006 I started the job that has probably been the most enjoyable and fulfilling of my career. My first trip as Director actually began before my official job start date. It was to Washington to attend the first Serious Games conference in the USA. Over the next five years that it took to acquire the SGI building, equip it, promote it and recruit the staff needed to turn into a via commercial operation, I made many such international trips, mostly to speak at conferences and evangelise about serious games and immersive technologies.

During this time period I witnessed ever increasing improvements in the sophistication of the in-flight entertainment and facilities, and currently I rank the Singapore Airlines A380 as one of the best as it gives not only a fantastic selection of entertainment options which includes films, music and video games, but it also has a comprehensive set of educational activities including language lessons. I used one flight from Singapore to London to brush up on my Russian prior to an upcoming conference in Moscow. This personalisation of in-flight entertainment, giving the customer control over their experience of travel is a relatively tiny step forward compared to advances in other areas that will find their way on to the airline systems of the future as the travel industry gets increasingly competitive and airlines seek to differentiate their journeys from their competitors.

Apart from the 1966 school trip cruise back from Leningrad via Helsinki and Copenhagen, a mini one day cruise to the Bahamas from Miami, my only other extended amount of time on water was a commission to speak on a commercial cruise organised by Richmond Events in 2010. The 3 day trip from Southampton to moor off the Channel Islands was designed to bring corporate buyers and sellers together, mainly at specially arranged meals, through a series of presentations and workshops run by motivational speakers or recognised experts. This cruise introduced me to the late Clive Gott who sadly died a few months after I watched his motivational presentation called "Vapour Trails in the Sky" and joined him at one of the dinner tables. He was an amazing character and an example of how ordinary people can achieve extraordinary results with their lives (see www.clivegott.com).

Whilst moving about on land and in the air has given me enormous pleasure over the years, I have not always had the happiest experience with water, mainly because it scares me. For some reason, the black depths of the ocean and the "cruel sea" bring a fear of drowning. Whether it was because my Uncle Bill was one of the survivors of the wartime sinking of the battleship HMS Ivanhoe and his death was eventually contributed to by the shrapnel wounds he suffered during this incident, I may never know, but despite much encouragement from my parents and many hours playing with my cousins in the muddy creeks of Frampton Marsh I didn't learn to swim until I was about ten years old, just before I went to Boston Grammar school. This defining moment came in those same muddy creeks as the tide was receding one summer Sunday afternoon. Despite this breakthrough I got into difficulties in Boston's old corporation baths that used to operate near the General Hospital and Boston Docks and I suffered the ignominy of being probably the only swimming captain in history to have to be pulled from the water because he couldn't swim the fifty yards of his Under 12s race.

Controlling water transportation also had its setbacks such the incident at the Chapel St Leonards boating lake when I fell in and I was terrified when, as a relatively small and lightweight chap, I was dragooned into acting as a cox for the school rowing team in the Boston Regatta in the early 1960s. I hadn't the nerve to tell them that I couldn't swim very well. Happily that experience ended without tragedy. In the 1970s I went on a camping holiday to the South of France and sampled for the first time the nudist beaches of St Tropez. Whilst we were there I got introduced to wind surfing by a lovely Dutch couple who were masters of the art. I need not go into the details of the kind of pain I inflicted on myself as my naked body repeatedly slipped against the hard surface of the board. As if all these incidents weren't enough to put me off, I took my young stepson Christopher on a water-scooter on the inland waterway near Miami. What I hadn't bargained for was that unlike my Lambretta Li 150 where it was natural to shut off the accelerator to slow down, with jet skis this cautionary action makes you lose any steering control as it shuts off the water jets that both control your speed and direction. The net result was that I managed to collide with a motor boat coming in the other direction and cracked my ribs in the process. Fortunately for all of us in the litigation society that is the USA there was no other damage to humans or machines.

Water transportation has however played an enormous role in mankind's history as the communications channel for ancient traders, the field for conflict resolution between nations in major naval battles and the early preferred method of transport at the start of the industrial revolution. With the rising costs of fuel and the inaccessibility of some parts of the world, canals and rivers do still play an important part in the economy either from commercial traffic or tourism.

Whilst I was in Thailand in 2011 for the 3rd Thaisim conference I had my first experience of an elephant ride and a meal on board a floating restaurant in Ayutthaya as it travelled down the river past ancient temples. It was during this trip that I had the surreal experience of singing a Karaoke rendition of Roy Orbison's "Oh Pretty Woman" to a backing track featuring traditional Thai musical instruments. This river trip also brought home to me that some of the emerging technologies we are taking for granted still have bugs in them. I had purchased a brand new

Ancient Thai temples from the floating restaurant at Ayutthaya

Lumix camera in the duty-free at Heathrow airport – primarily because it was one of the new generation of digital cameras to have GPS tracking embedded. This allows you to capture the location of any picture you take as well as the time the photo was taken. The Thaisim conference was the first time I had used this camera in anger and, as I skipped through the pictures I had taken, I noticed that the description of the location for the Thai temples was "Hospital for Venereal Diseases, Ayutthaya" and the pictures of my conference presentation were attributed to a local shopping mall instead of the college where it actually took place.

The technology of canal transport seems to have changed little over the last two centuries except for the evolution from horse drawn barges to the modern diesel powered canal boats that two of my closest friends, Steve and Claire Williams, live on at Blisworth in Northamptonshire where they manage the local marina. Claire entered into my life courtesy of a commuter train from London on which we were both passengers. Sat on the opposite side of the corridor on this train, I was "people-watching" Claire with fascination as she and a fellow passenger drank a bottle of white wine together, something I had never seen before on any

The author with Steve and Claire Williams and fellow narrow-boater Steve Howe at a joint birthday bash

of my rail journeys. I detected her South African accent and reached the conclusion that she and her male companion were very close friends so I was surprised when she stayed on the train at Kettering station where he alighted. It later transpired that he was a complete stranger she had just met on the journey. As the train stopped at Market Harborough and we both got off, I asked her if she was South African as, at that time, I had a member of staff who had just come to the UK from Johannesberg. Thus began a lifelong friendship and the beginning of many "Alfie" business lunches in which all participants actually network by telling the harsh truth about their business problems.

Claire and Steve met through internet dating, a subject I will discuss in more detail in a later chapter. This union through internet technology led to Claire joining Steve in a joint endeavour to fit out their own narrowboat "Penelope" from its raw metal hull state in the middle of a field in Oxfordshire. Today's modern wireless communications make it possible for Claire to work as a stenographer and transcriber from almost any location with wireless access and their home near Blisworth marina is witness to the renaissance of the waterways as a basis for both a domestic and commercial life. I was also fortunate at one time to live very close to Foxton in Leicestershire's historic flight of locks and its inclined plane used to haul bigger commercial barges up from the lower to upper levels and the site of a fascinating museum dedicated to the history of these British waterways. I used this history as an inspiration for the choice of a monologue in a special Old Tyme Musical Hall event at the Harborough Theatre. The monologue was written in the early part of the twentieth century as the canal network was in a terminal demise brought about by the railway network and was called "The Day I saved the barge" – a comic tale of a runaway barge towed by a former winner of the Grand National.

My only other enjoyable and notable encounter with water transport technology was an occasion when I visited the American Adventure theme park in Derbyshire. This theme park built on the former site of a coal mine attempts to mimic its USA counterparts set up by Disney and has a sizeable lake in its heart. The day that I went just happened to coincide with the visit of an iconic boat from the movie industry. The original "African Queen" of Humphrey Bogart and Kathryn Hepburn fame had been refurbished and converted from steam to diesel by its owner and was doing a tour of locations in the UK. I had the privileged experience of steering the very tiny craft that Bogart had towed through the jungle swamps and was interviewed by Radio Derbyshire in the process.

Of all the transport technologies that have influenced my life, it is probably road transport and travel by car that has seen the biggest revolution. The days of the hand-cranked, temperamental Humber Snipe in the 1950s in which drivers had to use all their skill, care and intuition to get the best from their machines have long gone. Today, highly automated and precise engineering brings us vehicles that are not only more reliable and in less need of human intelligence to operate to their full potential but can also be individually customised before manufacture. Purchasers of new cars can choose on-line from a whole range of colours, features and accessories and know that their car will be built specially for them within a relatively short timescale. Privileges that could only previously be afforded to the very wealthy are now available to the man in the street.

Whilst this in itself is an amazing development within my lifetime, it is the technology within the car itself that has seen the most profound changes in personalisation and artificial intelligence. Even within my modest Fiat 500 I have Bluetooth technology that includes speech

The author's Fiat 500 dashboard with Bluetooth and satellite

recognition, digital media player, hands-free telephony and satellite navigation. My car remembers all my contact names and telephone numbers and will dial a number for me by voice command. I can tell it to choose a particular track on my MP3 player and it will guide me to any destination in South East Europe by satellite navigation. My car is beginning to use its intelligence and superior memory capabilities to get the best out of me and it will not be too long before we see other in-car technologies that will extend this still further.

During my journey from my very first car, the second hand Triumph Herald Estate, through my very first new car, an orange Datsun Sunny, and the Vauxhalls, Renaults and Fords that I have owned over the last 40 years there have been remarkable changes and yet we are just at the frontier of these developments for many of the transport technologies, led and pioneered by the automotive industry and informed by all of the "immersive" technologies I will cover in this book. Thomas Cook would have been gratified that the revolution that he started in the 1840s has brought me and my fellow "baby boomers" so many rich lifetime experiences but he would be even more astounded at the unimaginable future that still lies ahead of us.

All of these experiences of transport technology as it has changed during my lifetime convince me that the days when humans used their skill, intelligence and sense of sight, sound, smell and touch to get the best from their cars, trains, boats and planes are fast becoming a fading memory of a long-gone time, only preserved in our leisure activities and our in-built and very human need to challenge ourselves. Today, as a result of the convergence of all the technologies covered in later chapters, it is the technology of machines which is using its intelligence and senses to get the best from us.

The journey continues but it may be better to travel hopefully than arrive at our final destination!!

CHAPTER THREE
Just Got to Get a Message to You

It would be wonderful to be able to go back in time to the early days of the Post Office into the era of the Penny Black postage stamp to appreciate how different social and commercial life was in those days. It is difficult to conceive now in these days of mobile phones, instant messaging, Twitter and Facebook how people ever managed to communicate in the days before the telephone and the internet. Some years ago, when I was involved in a community technology project in Market Harborough, I was given access to a local lady's collection of postcards from the early part of the Twentieth Century. Amongst these fascinating archives, which I set about digitally scanning for posterity, was some correspondence from the teacher in charge of the Harborough Grammar School girl's hockey team organising fixtures with schools in other towns. It was clear that the postal system of the day was a perfectly natural way to arrange such activities.

    I was born into an age when very few houses had telephones and most people relied on the public phone box not only for emergency calls but to communicate with loved ones at pre-arranged times. In those days there were two postal deliveries every day, one early in the morning and one around lunchtime. The post boxes were also emptied not once a day as most are now, but several times a day and it was not impossible to have a local letter posted early in the morning to be delivered the same day for the price of a first class stamp (less than 2p). Looking through a collection of my late father's wartime letters home to his sister, my Aunt Dorothy, was a revelation about their fond brother and sister relationship and about life in wartime England and in the RAF in India. I only wish I had seen these whilst my father was alive because I would have loved to have heard him tell his story. My Aunt Dorothy had the educational opportunities denied to my father because of the war and the early stages of her courtship to Craig Bedford who would later become her husband are beautifully articulated in this exchange of letters. As I grew up, I made active use of this messaging technology for love letters to my adolescent sweethearts, and to my parents on the occasions when I was away from home. The richness of this information has mostly been lost in the instant access to messaging technology we have today. Letters are a form of asynchronous communication which allows the writer to carefully choose the right words and the reader to think about and consider a carefully penned reply.

Even after I got married and had access to telephones, I continued to send weekly letters to my parents to bring them the news, and my dad would type out his reply on an old Remington electric typewriter I had given him to overcome the difficulties in reading my dad's appalling handwriting. I still have some of these exchanges of correspondence today and they bring to life my memories of the 1970s and the start of my career in telecommunications and media technology. When I was at school, I used to get part-time work at Xmas working at the Boston Sorting Office where my Uncle, Craig Bedford, was Assistant Head Postmaster. This was in the days before postcodes were introduced into the UK in 1967 and I can still recall how my knowledge of geography improved after having to allocate letters to pigeonholes or sacks for onward delivery. The early mechanised letter handling machines still required a human being to look at the address and type in a routing – human intelligence and eyesight getting the best out of a machine.

Today, textual communications have become much more synchronous and instant. People of all generations are using SMS text messaging to fire off feelings, thoughts and information in far less rich and considered ways than handwritten letters provided. Text sent this way is seen as much more convenient and effective than a voice telephone call which runs the risk of going to voicemail and not being responded to. Before SMS text, telegrams were the only way to get urgent messages from distant parts. I still remember from early films the sight of the telegraph operator tapping out messages in short sentences punctuated by the word STOP. In wartime, these brown envelopes with lines of black text on white strips of paper, and delivered by hand were a vital part of daily life.

The very first telephone I ever used was in an iconic red public telephone box with the instruction to "Press Button A" when your call is

Timeline of Media Technology Adoption

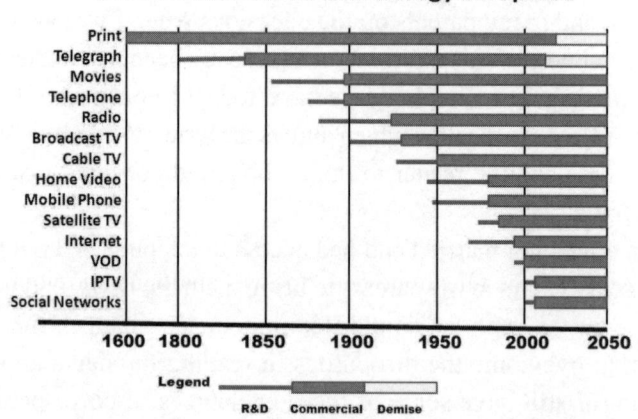

Technology Adoption Timeline (Wordpress.com)

connected. Calls cost 4d (old pennies) or just over 1p in today's decimal currency. Sadly this first ever call, made when I was at primary school, was unsuccessful as I dialled the wrong number and lost the money I had in my pocket, causing great panic at home because I had forgotten that I was not going to grandma's for lunch that day and was trying to ring my mum to tell her I hadn't got time to come home for lunch and so I went back to school instead. My frantic parents, expecting me for lunch, contacted the headmaster who hauled me out of class to admonish me for causing such worry to my mum and dad. We had lived in our council house in Wyberton for some years before we had the telephone line installed and our number was Boston 4040 – perhaps it was the trauma I caused through that wrongly dialled number that helped me to remember that number to this day.

I can barely remember either myself or my parents ever using our home phone – simply because very few of the people we knew had phones themselves. Like so many early telecommunications inventions, it took several years of research and development and many more years of commercialisation before these technologies became sufficiently widespread amongst homes and businesses for them to become a regular part of daily life. This phenomenon and today's more rapid adoption of any new digital communications technology is reflected in the technology adoption timeline chart shown above. The heavy Bakelite phones that began to find their way into homes used a rotary dial which sent a series of electrical pulses to the telephone exchange to control the old Strowger switches that made a direct point to point connection to your dialled number. Before the invention of Strowger switches, in the days of the "candlestick" phones, users would crank a handle to alert an operator on a big switchboard to manually plug your line into a socket on the board to make your connection.

Unlike today, the telephone network was analogue and your voice calls were routed entirely through copper cables which went from your telephone exchange via trunk cables to the green distribution cabinets you still see beside the road, and onto lower capacity cables to telegraph poles known as D.P.s (Distribution Points). Essentially, unless you had what was known in those days as "shared service" in which you and a neighbour shared a telephone line, you had your own dedicated pair of copper wires from the phone in your house to the telephone exchange. Your voice would be transmitted via a crude microphone and its signal amplified and distorted as it made its journey across the network.

My first introduction to the digital world and telecommunications that became my career was in 1966 went I picked up a book on digital technology at Boston railway station on my way to London for the interview for a Post Office scholarship scheme. The reasons I had chosen this as a potential career option were quite simple and very mercenary. I believed that telecommunications must be a growth area offering good potential for career progression and the GPO scheme paid more money to students than any other scheme I looked at. As my parents were not well off, this seemed like a win-win solution to getting to university. As the book I bought explained very simply, digital signals worked on a binary "on or off", "1 or 0" basis which was ideal for computers but also meant that our analogue voice signals could also be converted to a stream of binary digits or bits, bringing major opportunities to change the shape of our telephone network. At my interview, the mysterious new "System X" telephone exchanges were mentioned and I was asked for my views on introducing new technologies like digital and whatever my answer was did the trick because I won a scholarship.

My initial week long induction course for this scholarship began in

London and it was then that I spent my first night in a proper hotel in Russell Square not far from Kings Cross. My only memory of this stay was that I was very nervous going down for breakfast and, in my rush to leave the table, the back legs of my chair got stuck in the deep pile carpets and as I tipped backwards my knees lifted the table sufficiently to spill cereal into the lap of the man opposite. The first six months of this pre-university year was spent on a variety of courses including the workshop practices course in Bootle, technical courses at Dollis Hill and a number of courses at the rural mansion that is Horwood House located near what is now Milton Keynes. These courses at Horwood House tested my boredom threshold and almost led to me losing my scholarship. We were 3 weeks through a six week course and the novelty of the in-house bar had begun to wear off when I read in the paper that the BBC were not renewing Harry Corbett's contract for the "Sooty and Sweep" puppet shows. I drafted a letter complaining about this heartless and unwarranted decision and suggested that one of our student apprentice number might one day become Postmaster General and refuse to renew the BBC's license, setting up instead a TV channel to broadcast Sooty and Sweep every day. I persuaded everyone on the course to sign their name in a circle at the bottom of the letter and duly posted it to the "Points of View" programme hosted at that time by Robert Robinson. My biggest mistake in this foolish and mischievous action was to use headed notepaper from the Horwood House Training College.

The morning after the Points of View programme was broadcast (still in black and white) – which none of our group actually saw – our resident tutor Alf Stokes came into the classroom with a very sombre face. Alf, whose nickname was "Or is it ?" after his habitual response to any technical statement he wanted us to ponder, was a portly man with a normally jolly persona. When I saw his flushed and sombre face I began to fear the worst and when he said "Who is responsible for that letter on Points of View last night? We have been having phone calls from Post Office Headquarters all morning" my life began to pass before my eyes as I contemplated the shame of being sacked for such a puerile act. In true George Washington style I owned up at once and there was a horrible silence and a deliberately extended tension which Chris Tarrant on "Who wants to be a Millionaire" or the host of "X Factor" would have been proud of before a smile broke across Alf's face as he said "Fortunately the bosses took it in the spirit in which it was meant and thought it was very funny" In that instant I had gone from Zero to Hero but I never sought to repeat the exercise.

Much of the second half of the pre-university year was spent in my local telephone area whose headquarters were at Peterborough. I did a small amount of training in Peterborough itself but most of my time I was visiting the different teams who did the work at the telecommunications "coal face" in and around the town of Boston. This included shadowing various technicians doing internal work in the telephone exchange where the Strowger switches were constantly clicking and whirring, time spent in the planning teams, customer sales processing and finally with a telephone installer called Brett. This was the most fun of all because Brett's "territorial patch" included the seaside resort of Skegness and the villages north of Boston. Brett was a lively character who regaled me with anecdotes worthy of a "Confessions of" film which were very popular at the time and often starred a young actor called Robin Askew. During this time I also worked some evenings at the Star Inn at Sibsey behind the bar wearing a white shirt and bootlace tie to fit in with the cowboy saloon ambience.

Once the pre-university year came to a close it was time to start my Electronic and Electrical Engineering course at the University of Birmingham. Although I was involved in some work with computer programming I cannot recall anything on the course dedicated to telecommunications. The two fellow student apprentices who came to Birmingham on the same course, "Jim" Ames and Don Fisher, were as different from each other and from me as chalk and cheese. Both were more technically and academically able than I was and Jim was especially studious and bright. Don, whose family lived in South London at the time, originally came from Middlesborough and he had a love of alcohol, especially Newcastle Brown. All the new undergraduates were encouraged to join their faculty society during "freshers week" and so Don and I signed up for the Electrical Engineering Society and were duly conscripted for the "boat race" initiation ceremony that consisted of 2 teams of four people, each of whom had 2 full pint glasses of ale. The boat race took the form of a relay in which the first person in line drank one of his pints and put the empty glass upside down on his head as a signal for the next person to repeat the trick and so on to the end of the line. Don happened to be in the most challenging position as the last man. This meant that he had to drink one pint very quickly, put the empty glass on his head, before starting immediately on his second pint. Unadvisedly, Don had already drunk more than one pint of his favourite Newcastle Brown during the evening and instead of tipping the empty glass upside down on his head, he downed the first pint and tipped the second still brimming glass upside down on his head. Don began to break into laughter, along with everyone else, at this turn of events but his laughter quickly changed to an impression of a beer pump as he vomited the first pint back into the empty glass without spilling a drop. In our inebriated state we were lost in admiration but I am glad to say that such drunken episodes were few and far between.

In the university years from 1968 to 1971, most of my messaging was done via weekly letters home although in my first year I regularly came home at weekends to play football for my Sunday League team, Gipsy Bridge. It wasn't until I finished my final exam and scraped through my Honours Degree with a BSc (Hons) 2.2 grade that I began my career in telecommunications in earnest. If there is one thing that I had learned during these three years in Birmingham it was that I was far more interested in people than in research and development which, in essence, is what all we student apprentices were being groomed for. When it came to the Post Office placement interviews I made my preferences known and to my surprise they took me at my word and offered me a middle management position of Executive Engineer in the Leicester Telephone Area. My mum was initially dismayed at this job title because the word engineer to her meant that I would spend my days wielding an oily rag. When I arrived in Leicester I was interviewed by Mr Astle, the Leicester Telephone Area Manager at board level in charge of the Sales and Installation teams.

Mr Astle was a long-serving servant of the GPO who had spent his life's career in the organisation to reach the lofty heights of Area Manager in his 50s. You can imagine his disdain at the prospect of having an enthusiastic but naive beatle-haired and bandido-moustached twenty two year old straight from university reporting to him. The ensuing events led me to be certain that he didn't find me suitable to lead a hardened team of installers and so it was that I was allocated to Mr Pikett, Area Manager of Internal Works, to take charge of EI7 Internal Planning Team. This was a motley crew which included the Radio interference team and staff based in the Leicester GMO. My office was marginally bigger than a telephone box and I was allocated a mentor in the shape of Dave Salmon, an Executive Engineer whose reputation for practical jokes was legendary and went before him. Under his tender eye I had responsibility for around 21 staff with a relatively modest budget.

The Radio Interference (RI) team were something of a legacy of civil service days. My Assistant Executive Engineer in charge of this area was John Smith and my practical joking mentor Dave Salmon told me to watch out for John's dodgy expenses claims which, in my naivity, I discussed with John to my great embarrassment and Dave Salmon's enjoyment. The RI team would investigate radio interference complaints from members of the public which could include pirate radio broadcasting. I recall one letter from a disturbed man who claimed someone was broadcasting radio waves which were reading his mind and spying on him which proved quite a challenge to draft a formal reply to, but today might not seem so far-fetched.

I began this new career challenge in enthusiastic fashion by asking each of my staff to come in for an individual chat. Whether this in hindsight was a mistake or not remains to be seen but my interviews included one with a member of staff who had an unfortunate "tick" when he was nervous. During our friendly informal chat in my telephone box of an office he would periodically throw his head to one side and sing "Doodle ip doodle doodle ip" in a high pitched tone. I found this most disconcerting to say the least but we both struggled on pretending not to notice until the allocated 20 minutes was up. I should have realised at this point that EI7 was the "Dirty Dozen" of the Leicester Telephone Area and that I was expected to be their Lee Marvin.

During the first months of my tenure in this post I went on various courses to bring my specific knowledge for this role up to speed. The profound changes from analogue to digital were just beginning to happen. The old Strowger telephone exchanges were being replaced by ones where the calls were routed through switches that used mechanical relays that were either "on or off" and as time progressed I witnessed the introduction of PCM (Pulse Code Modulation) and TDM (Time Division Multiplexing) into the telephone Network. Until PCM and TDM, the way the Post Office used to carry several calls simultaneously down one pair of wires was to use Frequency Division Multiplexing, similar in technique to the way radio channels are separated by using different frequencies. PCM and TDM were truly digital techniques and the GPO, which soon separated the postal and telecommunications businesses to become the British Telecommunications (BT) that is familiar today, began a programme of installing PCM relay stations at regular points in the network.

PCM relay stations were used to "clean-up" the signals, amplify them and relay this cleaned-up signal to the next relay point. This cleaning up was a simple process that looked at the incoming signal to see if it was a 1 or a 0 and then send a pristine and amplified version of its decision onto the next PCM station. TDM worked by allocating slots in time for each telephone call instead of slots in frequency and it was through these embryonic technologies that packet switching came to be introduced. Packet switching is still somewhat mind-boggling to most people (including myself). Instead of your telephone call having a completely dedicated pair of wires from your phone to the person you are calling, your analogue speech is converted into digital information arranged in small packets of data that contain not only your voice but also routing information about who you are connected with. This technique allows the network to send these packets of data through different routes to the same destination based on what cables and switches have spare capacity, and then re-assembles your call into its correct order before it travels along the local network to its destination.

Text messages were somewhat simpler to handle because text characters and numbers are easy to represent in digital format. The telex network of those days was an efficient way for businesses to quickly send information from point to point and the messages would be printed out on mechanical devices that operated like a typewriter. These "teleprinters" were used for many years to bring television viewers that latest football scores on a Saturday afternoon. The TV cameras would be trained on the print head and millions of people would watch it chatter out each score as it came in. The football commentator reports of finished matches would also be done by telephone, often with a black and white photo of the reporter describing how the match went.

My first few years in the Leicester Telephone Area were punctuated by another scholarship award for a one year full time postgraduate Diploma in Management studies (DMS) at Leicester Polytechnic, later to become De Montfort University. We were based in the James Went building, now long demolished and the site of very recently erected new faculty buildings for Law and the Performing Arts. The James Went building was best remembered for its paternoster lift, a bizarre contraption that operated like a lift without doors that you jumped into as it went past your floor. I doubt whether such devices still exist today, if only because of the health and safety issues it must raise. I came away from this DMS course with a distinction, probably because of the final project – a statistical analysis of residential telephone usage at the different exchanges in the Leicester Telephone Area. The statistically different usage patterns from random samples of 40 subscribers at each exchange led me to the conclusion that the biggest users were in new towns and village developments where the inhabitants would have migrated from their home towns and villages and consequently made more phone calls to keep in touch with the friends and family they left behind. This information made an ideal platform for the "Buzby" marketing campaign which BT launched in the late 1970s, targeting the calls to friends and family in distant towns.

This spell in the Leicester Telephone Area with what changed from Post Office Telecommunications, part of the old GPO with its civil service rules, to the autonomous British Telecom, lasted until 1978. During this time I moved from Internal Planning to External Planning and then on to External Works where I had responsibility for around 200 staff and a substantial budget. During this period I began playing soccer for Hinckley Athletic in the West Midlands league. I suffered an early season injury which ruined my chances of playing for the National Post Office team though I did captain a Midlands Post Office side which beat the National team in a friendly fixture. I also took up playing squash against Post Office friends at the Wanlip Squash Club and my squash playing days became more serious after a knee injury sustained whilst playing for Hinckley put paid to my soccer ambitions. I later moved on to Squash Leicester, a brand new club built beside Leicester Racecourse and owned by two former County Squash players, Ian Turley and Richard Wilson. Not only did Squash Leicester become the focal point of sports life but it was the place where Rentacrowd came into existence with many fantastic nights at the club. During this period I used to get a fortnightly request from a friend called Rick Taylor to play against his son Simon who was fourteen years old and showed a lot of promise. I was a first team player at the club by this time and good at controlling the squash ball. This regular games against Simon continued over a period of several months until he eventually beat me on one occasion, after which the phone calls from Rick stopped as Simon moved on to challenge better players, eventually becoming the English Number One player.

Post Office Telecommunications provided me with a good social as well as professional life in the Leicester community. Together with Diann Lester and Marian Hughes, who was the General Manager's PA, we set up the Post Office Squash Club which started at Wanlip squash club north of Leicester and, for a number months when I was the only one of the three of us with a proper membership there – a husband and wife membership with my wife Angela, both Diann and Marian masqueraded as Mrs Wortley on more than one occasion. Diann is married to Malcolm Lester and they have remained lifelong friends with whom I shared many happy times. Malcolm was a good singer and he and his now sadly late friend Ian did a very good Blues Brothers tribute act.

In 1978, after four years in the Executive Engineer grade, I became frustrated at the seeming lack of promotion opportunities. I had excellent job appraisals but because my "seniority number" was outside the range of those being interviewed for Area Manager jobs, I wasn't deemed ready for promotion. Impatience got the better of me as I sought a way round this seemingly intractable problem. I discovered that the Post Office were planning to rationalise their management training facilities into one central college to be based at Newbold Revell near Rugby. This was within commuting distance of my home in Great Glen and gave me what I thought was a back door into the next level of management. I reasoned that when this new college opened, there would be vacancies for Senior

Management Tutors, a higher grade than mine but not bound by the same promotion restrictions. My plan was to get a job as a Management Tutor at the existing college at Cooden Beach near Bexhill, work there for the anticipated 9 months before the new college opened, and then apply for one of the senior posts that was bound to be created.

Like many of my ingenious schemes it didn't go to plan. I secured the Management Tutor post at Cooden Beach, went to be trained on how to teach at a course in London, and then began nine months of a 150 mile weekly commute down from Leicester to Bexhill. The route through London made the suburbs of Finchley and Hampstead very familiar to me as my journey took me through Archway, down the Holloway Road to the Blackwall tunnel under the Thames and through the East Sussex countryside. I stayed in a Bed and Breakfast near Cooden Beach and had a regular routine of walking along the beach in the mornings. The course I was teaching was to newly promoted first line supervisors and was mostly about man management techniques. There was some existing material to work with but I was given a lot of freedom to develop my own ideas. It was here that I first came across the concept of a "serious game" in an existing workshop called "Action Maze".

Action Maze consisted of a box of cards describing typical work scenarios around a problem with a member of staff. It began with the first card that told you about the starting situation in which a previously reliable and well-liked member of staff suddenly started coming in late and making a series of weak excuses. From this scenario you were given a series of decisions to choose from including doing nothing, disciplining him, counselling him or just letting him have an arranged later start. These choices were meant to reflect the conflicting interests of job needs, team needs and individual needs. Once the choice was made, this led to a new situation where once again you had different choices to make and so on through about 3 levels after which you reached an end-point where you had either successfully balanced these conflicting needs or you had caused more problems for yourself through making the wrong choices. This kind of hierarchical decision tree is still used in modern day electronic serious games.

The challenge of this new career opening made the Monday morning and Friday afternoon round trips to Bexhill, coupled with the peace and calm of the East Sussex location of the Post Office Training College in its mansion house overlooking the putting green on its front lawn, almost enjoyable, especially in the Spring and Summer. As the months wore on it became increasingly obvious that the transfer of all Post Office management training to Newbold Revell was running into problems even though the property, a former girl's boarding school with massive grounds, had already been purchased for a substantial sum. Union opposition had caused a change of heart and these plans were put on ice, if not completely forgotten. This left me with the choice of either relocating to Bexhill or trying to get a position back in the Leicester Telephone Area. By this time I had completed and implemented all my original course innovations, including a silent video I made to test students' interpretation of body language in interview situations, and I found myself with large chunks of free time in between student facing sessions. The original cunning plan to accelerate my career had misfired and my low boredom threshold led me to think hard about myself and what I would like to do with my life. In the event, it was during the early part of 1979 that I made the strategic decision to leave the Post Office and seek out a career where the rewards were more in proportion to your abilities and effort.

My entrepreneurial streak recognised the opportunities that the convergence of telecommunications and computing technologies would create and I decided that I would like to set up my own business in this sector. The bureaucratic and legacy civil service mentality of the Post Office at the time meant that I needed to bridge my skills gap with some form of experience in computing, business enterprise and sales and marketing before I could feel ready to take this substantial new step in the right direction. The answer that came to mind was a job in a large company selling computers. IBM, NCR and Burroughs were all big names in the computing industry and I had seen sales jobs advertised for all of them. I applied to IBM and NCR and got shortlisted with both companies. The NCR interview was carried out by someone I knew socially from my local drama group at Kibworth, an avid Scottish golfer called Ian Sutherland. It became clear from the interview that NCR needed sales people to hit the ground running with both computing and practical commercial business experience. IBM on the other hand offered a full year's training before you were qualified to go out and sell, and their reputation for sales training was second to none. I joined IBM in 1979 as a stepping stone to having my own business and succeeded in this cunning plan when I left in 1984 to set up my Mass Mitec consultancy in Leicester. My time in IBM is covered in more depth in the chapter on computing technology but it was with IBM that I first encountered the use of instant messaging between computer terminals on a network.

It was also my time in IBM that gave me the idea that catalysed my forming my own consultancy in 1984, Mass Mitec (Marketing Administration Support Systems – Marketing Information Technology Consultancy). The idea came from a case study I learnt about from one of my colleagues. Investment in computers in this era had almost entirely been based on cost savings from staff savings and/or productivity improvements. Nearly all the solutions I had in my kitbag were intended to deliver a return on investment this way. It was the pharmaceutical distribution company Unichem's use of telecommunications technologies for competitive advantage based on innovative new customer-focused services that made me detect a sea change in thinking based around technology for marketing led innovation. Unichem had stolen a march on their competitors by offering their customers a win-win proposition which entailed them buying a hand-held terminal with a modem cradle for £5,000. Under this arrangement, the chemist could use the hand held device to scan the bar codes of all the drugs he needed to re-order and after scanning all his required drugs, he would place the terminal in its modem cradle and the system would automatically dial Unichem and place the order for replenishment on the next delivery. This elegant and innovative solution meant that the chemist would normally get his drugs within 24 hours and the arrangement was such that the chemist could opt out at any time and Unichem would pay him back the original £5,000 and collect the terminal from him. This idea caught my imagination and as the IBM PC had just been launched I figured that small companies like mine could deliver big company services in similar innovative ways.

Mass Mitec started life in the offices of Venture Business Forms above an opticians in London Road, Oadby, Leicester, facilitated by a deal I did with Venture's owner and Rentacrowd friend John Porter. In

Mass Mitec's first office above the Opticians on London Road Oadby

exchange for getting his business a second hand IBM System/34 mini computer, setting it up and looking after it on a day to day basis, he gave me £6,000 start up capital and free accommodation and secretarial support for one year. The £6,000 I immediately invested in the technology which I believed would give me competitive advantage, setting myself up as a Living Laboratory to prove that my business concept would work even in the smallest micro-enterprise. My 64k twin floppy disk, monochrome display IBM PC cost over £3,000 and I supplemented this with a Mannesmann Tally dot matrix printer costing over £500. My earliest software and hardware acquisitions included Multimate Word Processing software and a device called the Braid Telex Manager which turned my humble desktop computer into a telex machine, enabling me to impress potential clients with telex messages drafted in Multimate and sent over the same telephone line as my phone.

Mobile phones and wireless networks were still a pipe dream in those days. I wasn't until the 1990s that I got my very first mobile phone, a huge "brick" of a phone made by Motorola. This was the "must-have" accessory for any budding executive and weight lifting one of these phones from your briefcase was a sign that you were a serious player. I was actually able to make a phone call with my Motorola on a train from

London to Leicester in those days, something that today is practically impossible on the East Midlands Mainline. When I recently challenged a train manager about the non-existent mobile phone signal she blamed the train widows for "having the wrong kind of glass" and suggested I stand in the corridor between carriages to overcome this problem.

One of the disciplines I started in my time with IBM was the keeping of electronic records of my marketing database. I had developed my own system in IBM using a System/34 minicomputer programmed in its RPG language and once I started Mass Mitec I chose a database package called RBase to set up my own system with details of contact names, addresses and telephone and telex numbers. At this time there was either no such thing as a facsimile (fax) machine or there were very, very few companies using them. Multimate had a mailmerge function which I used to send out telex mailshots to named contacts in my marketing database, a technique replaced in later years with e-mail newsletters. I would also use the mailmerge and my dot matrix printer to generate letters to these contacts which were posted by hand. This ability, very innovative for 1984/5, helped me to get blue chip companies interested in my services and senior managers to attend my regular seminars on technology held in local hotels or on site.

One of the clients I was able to acquire as a result of contacts from my IBM days was the textile giant Courtaulds. They provided me with regular consultancy work at their London HQ, Henrietta House, just off Oxford Street. I would charge £250 a day to set up and look after similar communications systems to the ones I was using in my Leicester office. I can probably also claim credit for introducing Courtaulds to their very first email solution for the desktop PC, based on Braid Systems innovative Mail Manager. Unlike the server based email systems of today, the Braid Mail Manager was a point to point system in which the sending PC dialled up the recipient's PC over a normal telephone line and transmitted the message. I set this system up around twenty five years ago and it was a sign of things to come.

After a year based at Venture Business Forms I was ready to move on to bigger things and rented an office on the first floor of a building almost exactly opposite Venture Business Forms on London Road, Oadby. These offices used to be a hairdresser business and were located

Mass Mitec's first floor offices at 52B London Road Oadby

above a motor bike shop. Apart from having electricity sockets half way up the wall as a legacy of its hairdressing origins, it was a perfect size for an expanding company and I had my own office !! There was a main office big enough for up to about 10 staff, a storage room and kitchen facilities. Its address was 52B London Road and access was through a side door and up some less than impressive stairs. I can recall many initial visits from senior managers of corporate organisations where you could almost hear their footsteps slow down on the stairs as they wondered what sort of company Mass Mitec could be, operating from such an unprestigious entrance. Once inside, they were quickly persuaded that Mass Mitec was worth doing business with.

The other factor which made these offices so perfect was that I had just separated from my first wife and had moved into a two roomed cottage in Meadow View just 50 yards from my office door. The innovative use of emerging and often "bleeding-edge" technologies was an essential part of Mass Mitec's business development strategy. By the end of 1985, after we had moved into our new offices at 52B London Road, it had become clear that our corporate clients of the ilk of BT, Courtaulds, Boots, Fisons Pharmaceuticals and many others were most interested in buying communications technology and value added

JUST GOT TO GET A MESSAGE TO YOU 73

services from us, centred around our experience and use of business presentation graphics software and hardware. The early computer modems were snail-like by today's standards, operating initially at speeds of 600 bits per second. There were no reliable "plug and play" communications software packages and I personally spent a lot of time just getting things to work at all, let alone reliably.

These restrictions in data communications speeds led to the introduction of the "Prestel" 1200/75 standard in which data was sent at 75 bits per second and received at the princely speed of 1200 bits per second. This asynchronous data transmission system was ideal for the Prestel information services – coloured text on a black background, because the data you needed to send was very small, often just a page number, whilst the information received was a page full of text which resembled the later Ceefax service. By the time we moved offices again three years later I had acquired a new Amstrad portable computer which I dedicated to the Prestel service – which you could argue was a forerunner to the internet information pages of today.

My Meadow View Cottage

Amstrad portable PC similar the one I used for Prestel services

As the modem speeds became faster, the Hayes 2400 modem became a standard for data transfer although most communications software was unreliable and difficult to use, especially in the days before the graphical user interfaces from Apple and Microsoft Windows were introduced. This use of modems for the business stimulated a demand for Mass Mitec to sell modems and communications software. It was through this that I encountered for the first time a remarkable phenomenon that came to be a familiar pattern. In searching for a low cost but reliable communications software package I came across a solution called "Transend" developed and sold by a small business north of London. I wanted to negotiate a deal to distribute the software so I arranged to visit the developer at the house from which he ran his business. It came of something of a shock at the time to discover that he was a famous oboe player with the London Philharmonic orchestra and that he had developed this software in his bedroom as a kind of hobby. I had expected to find a geeky and technical person but this man, Malcolm Messiter, was the son of Ian Messiter who devised two of my favourite radio programmes of all time, "Just a minute" and "Mornington Crescent", both aired on Radio Four. Malcolm's wife was a concert violinist, also with the London Philharmonic and we concluded our software agreement with him giving me his solo "Oboe Fantasia" album on a cassette tape. Over the years I have seen that individuals with a creative arts background and no technical background have almost invariably been the most innovative in pushing back the boundaries of technology.

As telecommunications technologies and the telephone network improved, Hayes introduced a modem that could synchronously transmit and receive at 9600 bits per second and we began to use these in the company to transfer files and even do remote diagnosis and control of our client's desktop computers through programs like "PC Anywhere". Mass Mitec's core business had by 1988 evolved into specialised

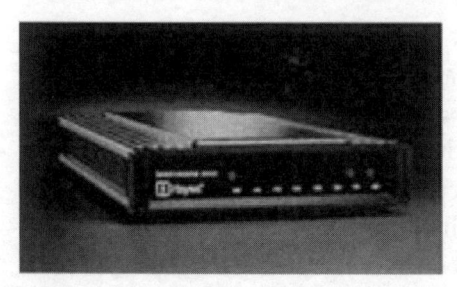

Hayes 2400 external smartmodem

*The White House Lubenham
– Home of Mass Mitec
1988–2005*

presentation graphics and we offered presentation design and imaging services to create professional quality slides and overhead transparencies for our clients. We were using our skill and intelligence to create compelling presentation content with software like 35mm Express, Powerpoint, Freelance and Harvard Graphics on the old DOS (Disk Operating System) PCs.

It was in 1988, as the three year lease on our offices in Oadby was about to expire, that I got the opportunity to fulfil a long-held ambition to own a property in the country big enough for both a home and a business. Susie Wiggington, my future second wife Carol's sister, worked in an Estate Agents in Market Harborough and knew that I was looking for a suitable property to buy. It was then that The White House in the village of Lubenham unexpectedly came back onto the market when a buyer backed out on the day contracts were due to be exchanged. The Howshams, an elderly couple who owned the property and were looking to downsize, were devastated so Susie gave us the opportunity to look at the property the same day and make an offer. I took my 2 key staff members Rachel and Gary to look at what had been the original home farm of Lubenham but was now reduced to a couple of barns, large courtyard and big farmhouse. With Rachel and Gary's enthusiastic blessing I had no hesitation in immediately offering the £110,000 asking price. So began another chapter in my life and the ongoing evolution of communications technologies. The history of the White House and the changes we made during our stay there could easily provide enough material to write a book in their own right. Having a successful business helped to secure the property and a £70,000 loan to redevelop the offices but our first hurdle was getting planning permission to convert the barn to the offices which would house Mass Mitec. The main problem was that the entrance to our courtyard was less than 80 yards from a blind corner on what was then, in the days before the A14, a main East-West route and carried quite a lot of traffic, including heavy lorries. A party of myself, my mother and my girlfriend met with the Planning Officer as he made his assessment of the potential dangers of using our courtyard as a car park. I was playing down the situation by focussing on our small staff numbers and low need to go in and out whilst my mother betrayed her pride in my success by telling him that we had lots of important visitors at Mass Mitec. Fortunately we won the day and began the job of turning this lovely property into a high-tech engine of rural development.

The main barn was a two storey building whose ground floor had once been a cattle shed and in later years had acted as a garage for the Howsham's car. The first floor was a disused chicken loft still semi derelict and covered in chicken wire and bird droppings. We engaged a local builder, Roy Langham, to carry out the conversion. Roy had also been long-interested in buying the White House himself and branded himself as a traditional craftsman, including using a carefully-restored vintage lorry for his business. Roy also lived in Lubenham less than a quarter of a mile from the White House. Our ambition was to equip our offices with the very latest communications infrastructure from the very beginning and to make ourselves as future proof as possible. A lot of the dirty work was done at the weekends by myself and my girlfriend and wife to be Carol Blake. This included clearing all the chicken wire and bird droppings from the first floor and demolishing an old greenhouse the stood at the foot of where we wanted an external staircase.

The converted barn that became home to Mass Mitec in Lubenham

The Howsham's old garage became our reception and what had been a cold-room in the days before refrigerators was turned into a small office and had a new set of stairs put in to make the loft above into my private office. The cowshed with its sloping floor and gulley for handling slurry waste was made into a big main office, toilets and a kitchen were installed and a small room was created to house a dark-room for processing the 35mm slides for business presentations. The upper floor was set aside as a seminar room for our planned series of workshops. All the old brick walls were "dry lined" but not before as many power and telephone sockets as we could envisage were installed and all the telecommunications and audio visual cabling concealed behind the panels that became the wall surface. We even pre-installed a television aerial and cabling to be able to access TV communications in the days before satellite TV. At this time we had a core team of six staff including myself but we installed the best PBX telephone system we could afford to give us four main external lines and up to a dozen extensions. We also set up dedicated lines for a fax machine and for data communications. All this work was completed during the autumn/winter of 1988 and we began our operations in full swing.

It wasn't until 1992 that the next significant development to totally change our business strategy came along in the shape of a British invention called Diskfax. This clever device did exactly what its name implies – it copied the contents of a disk in one place to another Diskfax in another location over an ordinary telephone line. Diskfax was housed in a rectangular black box with a numeric keypad on top, twin floppy disk drives for 5.25 and 3.5 inch computer disks, a built in 10MB hard disk and an integrated 9600 bps modem connected to an external telephone cable. There were many innovative and important features about Diskfax including:

- An ability to handle IBM, Mac and Unix disk formats
- No requirement for a computer
- The fastest data transfer speed available at the time from desktop devices
- 100% guaranteed data integrity on transferred files
- Unattended operation

- Connection to standard phone lines that could be shared with a telephone or fax
- Recording of the origin of all received files and displayed on its screen.

The arrival of Diskfax, which was originally intended to help software companies distribute updates more reliably and easily, happened at a time when graphics software like Powerpoint and Harvard Graphics were beginning to become standards in the corporate world and more and more company managers were able to create their own professional presentations. What was beginning to become a threat to Mass Mitec's presentation design services now became an opportunity to create an entirely new business model that would open up new markets for us.

At our White House offices we already had some of the most advanced presentation imaging technology available and our own in-house darkroom for film processing, but one of the most important barriers to offering a bureau service was the distance from our clients who were based all over the UK. Moreover, it was our rural location and growing competition from small presentation graphics specialists that made us vulnerable because our nearest town, Market Harborough, had very little demand for presentation services, except for Golden Wonder Crisps who had their HQ about a mile from us. Diskfax was the key to Mass Mitec setting up the UKs first National Presentation Network in partnership with Prontaprint, the high street print and copy franchise network of retail outlets.

I reasoned that Diskfax gave us a way of setting up sales outlets all over the UK at no cost or risk to the business. To make this business model work I needed to find a partner in a related line of business to ours who might have the interest in and motivation and resources to support a joint business venture. Of all the High Street copy and print retail suppliers at the time, Prontaprint had by far the largest network with over 200 franchisees in the UK and even abroad. My trials with Diskfax convinced me that it was capable of being used by anyone with a telephone line with minimal training and after-sales support required. The more complete story of this initiative is covered in the Imaging Technology chapter, but once Prontaprint management had bought into the idea and allowed me to present and exhibit at the national annual conference for their franchisees, I began a program that set up over 100 sales outlets across the UK and this became our primary source of income.

My biggest regret at the success of this deal with Prontaprint was that my father was not alive to witness it. My dad died of a stroke after a short spell in hospital. He was a loving but somewhat introspective and undemonstrative father – a man of traditional family values who did his very best to give me the opportunities denied to him. He was most able to communicate through his letters where he could find the words to show his feelings. In the days before the electric typewriter found its way into our home, his words were penned in a script that was almost indecipherable to strangers to his appalling handwriting. Communications Technology certainly assisted his communications capabilities but it was his handwriting that gave me one of the most poignant moments of my life during the short time he was in hospital. On the day that he was taken to hospital my mum telephoned to let me know he had been taken in but assured me that it wasn't serious, trying not to worry me. I was glad that my wife persuaded me to travel back to Boston to see him after a business meeting that morning in Reading because that hospital visit was the last time I saw my father alive.

As I entered the ward where he was being monitored, his eyes lit up as soon as he saw me but he couldn't speak because his throat was paralysed from the effects of the stroke. He was also having difficulty swallowing but he was alert and very happy to see me. As I talked to him about happier times and our joint memories and what was happening in my business, he could only reply through little handwritten notes he was able to scribble. As I tried to decipher one of these short messages I joked with him "You know dad, if the Germans had an encryption code like your handwriting we would never have won the war." He tried to laugh and in that moment of shared intimacy between father and son it was as if we had really being saying the words that our masculine stiff upper lip had never been able to express – that we loved each other. It was just days later that I got an early morning call from the hospital to tell me that he had just died and asking me to get in touch with my mum to pass on the bad news. I was so grateful for that brief few moments together in which we made our peace with each other and expressed the bonds we both felt.

So it was that just a few months later as I drove back down the A1 from Prontaprint's HQ in Darlington, after sealing this very important strategic deal, that I thought of how proud my dad would have been and I had to pull into a lay-by as I choked up and sobbed my heart out.

One of the significant strategic effects that the National Presentation Network had on my business was that it reduced the dependence of the business on my skills and experience and created a "presentation services factory" that could be run on a day to day basis without my involvement and also created a motivated and trained sales force that I didn't need to pay a penny to in wages !! This reduced dependence on my involvement in the day to day operations had become all the more important because I had lost my two key members of staff, Rachel Smith and Gary Capewell, to a competitor. Rachel and Gary had been instrumental in helping me to build the business and their departure was a major loss.

Even during the euphoria of the growing Prontaprint driven imaging business I was conscious of the need for a long-term technology strategy that recognised the rapid growth in data projectors and the falling cost of desktop colour printers, both of which were a major threat to presentation imaging services which used 35mm slides and overhead colour transparencies. My plan was to help Prontaprint develop a strong High Street brand as a leading supplier of presentation services so that as bureau imaging of slides and overheads declined, their retail outlets would begin to sell presentation hardware and more specialised services that we would still be able to offer, and the business model would shift from a centralised service to become one of a distributor supplying to a retail network.

Once again, this was a cunning plan that didn't go as expected. Prontaprint had changed ownership more than once and each new owner brought with them their own agendas and strategic plans which, when Prontaprint was eventually acquired by an Irish company, completely derailed my plans because the new owners effectively wanted to copy my business model to serve their centralised large print facilities. By this time the data communications networks were getting much faster than Diskfax and the file sizes of presentations were just beginning to challenge the capacity of the floppy disk. Since around 75% of my business was generated through the National Presentation Network, this was a crisis for Mass Mitec and I had to effectively re-engineer the business over a short timescale.

It was during the latter half of the 1990s that the internet was beginning to catch hold and it was a robbery at my offices that ironically became a blessing in disguise. The Hayes modem speeds had also increased to a blistering 64K bits per second by this time and ISDN lines offered the capability to handle simultaneous voice, video and data. Although I had made my offices as future proof as I could from a telephone cabling perspective, I hadn't accounted for the growing development of local area network or LAN Ethernet cabling. Prior to this time, all of our office desktop PCs were stand-alone and we transferred files to each other by floppy disk. It was a robbery one autumn evening that changed all this.

That fateful day, I had been to visit my Uncle Len in hospital in Southampton where he was recovering from surgery that had developed complications. Len, like my mum, had smoked all his life and the arteries in his legs were beginning to get constricted to the extent that his feet were beginning to turn black. He was operated on at Winchester hospital where they replaced key arteries with some veins from his stomach region. His replacement veins worked perfectly but his bowels were accidentally nicked as he was being sewn up and he became very badly infected by this leakage which meant another operation to give him a colostomy which he has lived with ever since. This particular evening, as I drew into the courtyard at around dusk at 7pm, I was in a hurry to get to a rehearsal in the local drama group and, for once in my life, I just checked the office door was locked but didn't go into reception to see if I had any messages and check the alarm was set. Further alterations to our office as the business declined had reclaimed the room adjacent to reception for domestic use as a utility room and the connecting door had been boarded up. The house was fully occupied that evening but no-one heard or saw anything unusual. However, when I opened up the offices early the following morning, I saw some papers scattered on the floor at the end of the main office. My first reaction was to castigate my staff for leaving a mess the previous evening but as I came into the office I saw that all the computers had gone. The thieves had entered through a side window from School Lane that ran alongside our building. This window was well above head height and the thieves had forced open the security lock, dropped down into the office and, using some cutters, had quickly cut through all the monitor cables and taken only the system units. It turned out that one of our neighbouring businesses along Main Street had also been burgled the same evening. The police suggested it would be a professional gang from a different part of the UK using the motorway network to make a quick getaway.

I was devastated by this turn of events but all the staff responded well and we were able to continue operating whilst the Insurance Company was considering our claim. At one time I feared we would not get any compensation at all. I made a complete inventory with proof of purchase of all the stolen items and after what seemed an eternity in limbo, we were offered over £20k to cover our loss. With these funds I was able to re-equip the office and fit Ethernet cabling and servers to enable us to set up a LAN network and improve our data management and efficiency.

With the demise of our Prontaprint partnership, I began to focus on other emerging telecommunications technologies to give us competitive advantage and offer new products and services to our clients. The internet was beginning to be established and we made early forays into these applications by setting up a Compuserve email account. Later on we worked with a local Internet Service Provider called Web Leicester to host our web site and domain name www.massmitec.co.uk. I began to experiment with on-line collaboration technologies and desktop video conferencing. Intel entered into this market place around the late 1990s with a desktop solution called Proshare and a boardroom system called Teamstation. We used our ISDN line to connect an Intel Teamstation and began to sell systems in small quantities as I strove to re-engineer the company. The Intel Proshare solution was a network card you had to install in your computer along with a webcam, primitive by today's standards but able to handle simultaneous voice, data and video to allow document sharing, video and text chat and Powerpoint presentations in similar ways that are so commonplace today. The Teamstation was a boardroom system which had what was then a powerful Intel server at its heart, wireless keyboard and mouse and a good quality video camera that could do voice tracking.

The use of telecommunications technology for collaboration was just becoming an important part of our business when there was another defining strategic decision that involved the use of communications technology for competitive advantage. One of Mass Mitec's chronic and intractable problems had been our rural location and our remoteness from our target corporate clients, most of whom had city headquarters. I was reflecting on the growing issue of globalisation and the impact it was having on rural communities like my own. As the use of the internet expanded and transport infrastructure improved mobility, village stores, Post Offices and banks were all slowly disappearing and I thought about our own high tech business, able to attract blue chip customers from all over the world, yet seeing the bigger companies going to city-based businesses when they had the same, if not better services on their doorstep. It was the challenge the physical communities face in harnessing local skills, talents and services that led me to develop and submit an application for a national competition called MMDP that Government of the day had launched around 1997.

The proposition and reasoning behind this award scheme was quite simple. As a nation of "shopkeepers" with a very high proportion of SMEs (Small to Medium Enterprises) the UK was not adopting the kind of new media technologies needed to incubate and develop fast growth businesses, so the Government wanted to offer some "pump-priming" finance to help support innovative solutions based around multimedia and the internet. Fuelled by a passion to use technology not only to help my business but also my local community, and with an intuition that many communities have huge untapped and unrecognised local resources, I developed a funding application for one of these awards with a project I devised called ComKnet (Community Commerce and Knowledge Network). The rules of the award meant that applications needed to be based on a consortium with at least one large partner. The awards covered 50% of the project costs and partners were required to match the Govt money with either cash or services in kind. The logic behind these rules was that small businesses like my own could not afford to only be paid 50% for their services or provide equivalent cash, so the sums awarded to each consortium partner largely reflected the size of the company. In ComKnet's case it meant that Mass Mitec would only receive £50k out of a total budget of £250k for the two year scheme even though, as originators of the idea, and the main implementers of the project, we would be doing most of the work whilst the bigger partners who also stood to benefit from Comknet got larger sums !!

There were eight partners in our consortium, including the Leicester Mercury newspaper, Loughborough University, Robert Smyth School and a local Internet Service Provider based in Corby. The objective of the project was to use the internet, multimedia and collaboration technologies to seek out, identify and recruit local talent, especially innovative users of these relatively new technologies, and help to leverage their potential to develop our rural community's social and economic wealth. I had some time before submitting the proposal been involved in an international knowledge sharing project organised by a guru called George Por. This project used a collaboration platform called Caucus and was essentially a web forum designed to build a knowledge base. My plan was to develop a community portal web site that would be based around Market Harborough and would encompass the surrounding villages. The business model for the long-term sustainability of Comknet was based on the web design business and consultancy the project could generate for Mass Mitec and could become our new re-engineered core business.

As soon as the project was approved and the contracts were signed, I set to work by doing some internet searches to find local talent and began by typing "Market Harborough" into all the top search engines of the time. I was surprised and somewhat horrified to find a web site which appeared at or near the top of all the rankings. This site, called Bigfern (www.bigfern.co.uk), had already embodied many of the ideas that I wanted to develop and was effectively a community portal which was probably better designed than Mass Mitec could achieve at that time. Here was another cunning plan on the verge of being derailed as I faced the moral dilemma of whether to support a potential competitor on my doorstep. I did not hesitate long however before sending an email to the webmaster of Bigfern, telling them about Comknet and inviting the webmaster to contact me to discuss their involvement. What followed was a story that I have told and re-told on many occasions all over the world and was a classic example of how my intuition that every

Bigfern Webmaster Frank Bingley

community has huge reserves of untapped and unseen talent.

The webmaster arrived one Friday afternoon and as he walked into our offices I thought I recognised him, and so I greeted him with the words "I'm sure I know you from somewhere". His reply both stunned and delighted me. He said "Of course you do – I'm your milkman." I nearly fell off my chair at these words because here was I, a well-educated man with a lifetime's career in telecommunications and computing and a serial explorer of new technologies, being given a lesson by the man who delivered my milk every day, had no formal higher education and was entirely self-taught. It transpired that Frank had built his own computer and had taught himself to write HTML using the basic Windows text editor called Notepad. My first reaction was that Frank came from a different planet but, as I later discovered many times over and in all parts of the world, every community has many Frank Bingleys, ordinary people doing extraordinary things because of a passion.

So it was that Frank and I set out to journey together, an unlikely couple – a career technologist and a hobbyist milkman, accompanied by my little dog, a massive black Newfoundland called Floyd after my favourite rock band, about to create a real-life mirror of one of my all time favourite films, "The Wizard of Oz". This journey along our own Yellow Brick Road of the internet from our metaphorical Kansas, Market Harborough also recruited our own Scarecrow and Cowardly lion along the way in the shape of Steve Cockayne and Bob Bridges whom we met along the way and persuaded to join us in our journey to find "The Great Oz". There were mayors and munchkins along our path as well as a few wicked witches. The ComKnet project was an extraordinary chapter not only in my own life, but also in the history of my use of communications technologies. It brought some very welcome and positive publicity to

Mass Mitec and acted as a gateway to the many papers and presentations I have delivered at conferences all over the world.

Over the two years Comknet ran between 1998 and 2000 we recruited quite a large number of extraordinary people as social entrepreneurs to support our project and throw their weight behind a vision to make Market Harborough a focal point of yet another revolution to mirror that started by Thomas Cook over 150 years earlier. I used Thomas Cook as my explanation for what we were trying to achieve for the project. The internet was the New Millennium's equivalent to the railway network of Cook's day and we would use it as a way of bringing new horizons and new experiences to our citizens by making it attractive, affordable and accessible. This was the message that our public and private sector partners embraced and led to our decision to make a video about the project. Although, using the basic video editing software available at that time on high specification multimedia computers I could record and edit my own videos, I wanted to get something more professional so I placed a small advert in the Harborough Mail newspaper (also one of Comknet's partners) inviting local people to help with the project.

It was this tiny advert that attracted the attention of our "Scarecrow and Lion" in the form of Steve Cockayne and Bob Bridges. They were the only two people to respond to the advert and once again reinforced my growing belief in the wealth of local talent. Steve Cockayne offered to do the filming for the ComKnet video. He was a former senior cameraman with the BBC who had worked on several major household favourites such as "Doctor Who" and "Eastenders" where he filmed the original death scene of "Dirty Den". At the time of his reply he was doing part time lecturing at a college near London to help subsidise his setting up of his own local video company. Steve also wrote science fiction novels, his greatest passion and, as if that weren't enough to occupy him, he was also a musician in a local band. The remarkable thing was that I had never met Steve before despite the fact he lived in the same small village as me, less than 200 yards from my house.

Bob Bridges volunteered to do the post-production video editing for me and I soon learned that he was a special effects expert who has worked on all the "Harry Potter" films as well as many other

Blockbusters such as "Lord of the Rings". Bob's house was in Market Harborough, less than 2 miles from my front door. These two amazing talents had been discovered quite literally on my own doorstep and were willing to bring their abilities to support the project. The Comknet documentary was devised as a "vox pop" case study of local people from many walks of life giving their thoughts on the impact of the internet and on how Comknet could put Market Harborough on the map. As well as helping to drive my own project, I had replied to an advert for a consultant to support a community technology project in South Kilburn, North London and was given a commission to partner with another consultant, a Scotsman called Robert Campbell who had impressed the selection committee with his expertise and willingness to relocate to South Kilburn to manage the project. The South Kilburn "Innit" project was similar to Comknet but different in the sense that the funding was being used to set up a community technology hub in a former laundrette in a deprived urban area where they were to offer different media and internet courses to more disadvantaged groups.

We had already done a lot of filmed interviews in South Leicestershire and South Kilburn when one Friday, as I was driving in my car, I heard the newly appointed Government e-Envoy, Alex Allan, being interviewed about his job as an evangelist for information technology and what it could do for the local economy. Alex was waxing lyrical about the success of Amazon in selling music and books on-line and he compared Amazon's use of personalisation techniques to the experience he used to get in his local record store where the staff knew his preferences and were able to suggest new albums he might like. Unusually, and perhaps rather foolishly, Alex also gave his email address out at the end of the radio interview and I hastily stopped the car in a lay-by to scribble it down. The following morning I was in my office early as usual, even though it was a Saturday morning, and penned an email back to him saying that I had heard his interview and telling him something about the Comknet project. The point I was trying to make was that Amazon's success in global internet commerce was also contributing to lost sales in local book and record stores and that we in the UK needed to exploit similar technologies to stimulate local sales and intra-community electronic trading and knowledge sharing. I went back into my house next door for some breakfast and was amazed to find that the Govt e-Envoy had actually replied to me within the hour. He recognised my argument and eventually invited me to meet him in Downing Street to discuss ComKnet.

The making of the Comknet video also gave me the opportunity to ask him if he would be willing to record a piece for our video to emphasise the Govt's support for such initiatives. Because I had devised the video to cite Thomas Cook and Market Harborough as examples of communications technology bringing social and economic wealth and I wanted to make this a common theme in all the vox pop interviews, I asked Alex Allan if I could write a draft script for him to edit and shape in his own words. I sent the script to him by email and he readily agreed for myself and Steve Cockayne to interview him in his Downing Street offices. When we did the interview he used my draft script almost word for word and suggested I could make a living out of writing ministerial speeches.

It was the making of this video, combined with the embryonic emergence of web conferencing technology called eVideo, that introduced me to another opportunity to innovate with bleeding edge technologies. There was an international company specialising in board room video conferencing systems at the time. Their UK headquarters were in Slough and I learnt that they were about to do a European launch of one of the very first PC based web conferencing solutions that could simultaneously handle voice, video, text chat and Powerpoint presentations. This software had originally been developed in the USA and been used by the military for collaboration between senior officers in different locations.

I contacted the company and persuaded them to collaborate with me to mutual advantage. I suggested that I could help get them a global audience for a launch live webcast if they would provide me with free access to their technology and resources for the event. We agreed on a plan that would broadcast one of the world's very first global webcasts in a one hour programme which would begin with live video streaming from their studio in Slough of a short fifteen minute discussion, accompanied by presentation slides, between Robert Campbell and myself about our respective projects. This was to be followed by the first public screening of the 30 minute Comknet video streamed over the internet and concluded with live Questions and Answers and response to questions and comments submitted by text chat.

The company effectively dedicated their Atlantic private circuit cable link to the streaming servers in the USA for this one hour session and I gained so much publicity for this very first webcast that we had 300 participants from all over the world registered to be part of this historic initiative. I had arranged a VIP special seminar at our upstairs office at The White House in Lubenham to be hosted by a new member of staff, a South African called Dick van Aken, who was of a similar age to me and had substantial technology and commercial sales experience over there before the violence in Johannesberg forced him and his English wife to re-locate back to the UK and Market Harborough. As the event began in Slough, Robert and I were able to watch the PC monitor screen with our video, slides and text chat apparently working normally, blissfully unaware that the massive demand for the event and the bandwidth needed to handle 300 online participants was too much for eVideo to handle and so consequently the vast majority of people online had no video or audio, just presentation slides and text chat.

This publicity banana skin left my enthusiastic and proud local VIPs less than impressed with the event and the technology. However, as I have found throughout my career, every cloud in my life almost invariably has a silver lining. When I got back to our offices and was told the bad news, I looked at the recorded text chat between the 300 participants and saw all the collaboration and online help activities that the webinar's technical problems had generated, I realised how powerful online collaboration technologies such as web conferencing could be. The event also, despite its problems, introduced me to a whole set of new international contacts with an interest in these technologies and enabled me to collect all of their names and email addresses to follow up the event by sending every single delegate a special certificate with a map of the world and their names on it.

Two of those international delegates were to play important parts in my use of communications technologies. Bill Harris from Portland and John Hibbs from San Diego joined this Comknet initiative and John Hibbs in particular was one of the most active participants in text chat as delegates exchanged experiences with each other and tried to help out. The use of web conferencing technologies became a very important part of my life from that point onwards. It was around that time that I was also invited to be a guest speaker via an internet connection at a conference in Kyoto Japan. My contribution was one of a number international presentations organised by Kyoto University. This was still in the days before broadband technology was really commercially available. This presentation to students at Kyoto University included video and presentation slides and remarkably used an early version of programme Microsoft Net Meeting. So it was that at 2:00 am I connected to the University and began to make my presentation over what was, from memory, made using a 64K modem. To check whether this technology actually worked in real time, I used the PC on the desk next to me to dial up the webcam set up at the back of the lecture auditorium and, for the first and only time in my life, watched myself live in Japan with the slides almost perfectly synchronised with the ones being displayed on my Netmeeting computer.

It was not long after the original eVideo webcast / web conference that I got a called from a USA company called HP Media Solutions, part of Hewlett Packard. My name had been passed by Bill Harris to one of their managers, Ken Codeglia, as they were looking to appoint a freelance Project Manager for a new programme of virtual classroom training they were organising to train field staff from Agilent Technologies at offices all over Europe. Agilent are global providers of electronic test equipment for the telecommunications and other major industries. They had been part of Hewlett Packard until a separate division had been created in the 1990s. The contact with HP Media Solutions, created as it was by a flawed experiment with a new technology, became a bedrock for my business and generated significant income until the services were largely brought in-house by Agilent in 2010. Ken Codeglia began to train me on the use of HPs own Virtual Classroom (HPVC), one of the early pioneers of such technology which allowed the simultaneous transmission of presentation slides and two way text chat. Audio was handled through a separate teleconferencing system using the normal telephone network.

Through this introduction I became the European Project Manager for HP Media Solutions responsible for facilitating and moderating these monthly online training sessions designed to update the field sales force with new product information, competitive products and the general state of the business. Every month I would go online in the virtual classroom to train new presenters on how to use the system as well as giving tips on best presentation practices. I soon developed my own training package which was also delivered online in the virtual classroom. For at least two full days every month I would facilitate a whole day session which began with the European Vice President of Sales giving an update of monthly sales against target and motivating the staff with strategies being put into place to beat the competition. This was pretty much like the monthly sales meetings I had experienced in IBM except that all the sales staff were in different locations. There then followed a series of technical presentations and updates delivered by the Product Managers, most of whom were based in the USA.

Bill Harris was doing an equivalent job to me in the USA for their time zone and was one of the most professional, generous and charming men I have encountered in my business. Bill and I used to share duties on the Asian time zone and cover for each other when either was unavailable. As part of their value added services, HP Media Solutions would also record each session and make it available online for those who could not attend the live event. This involved Ken's team in the USA using special editing software to ensure the archived version was delivered as smoothly as possible. Over the next few years I and my South African employee Dick van Aken would facilitate many of such sessions including quite a detailed analysis of each session to identify how these online sessions could be improved.

It was during one of these monthly online sessions that a sad moment in history unfurled as the Twin Towers were demolished in the 9/11 terrorist attack on New York. Operating as I was from my main office, without access to television or radio, it was through text chat and audio comments from my presenters during the pre-briefing and post-briefing sessions that the true magnitude of this terrible moment in history became apparent. It began with just a few text comments about a plane accident in New York and grew into a realisation that it was no accident. It was a credit to HP Virtual Classroom technology and the telecommunications infrastructure supporting it that even though many telephone networks in the USA collapsed under the volume of calls and messages, the virtual training sessions barely suffered any delays in the synchronisation of the slides. Sometime later, these online training sessions were transferred from HP Media Solutions to Agilent themselves, largely I believe to avoid any commercial competition between Agilent and HP. Bill and I were both retained as external contractors during these changes, and the virtual classroom also changed from HPVC to Placeware and eventually to Microsoft Live Meeting.

Comknet had helped to bring a new business model based on the use of communications technology for competitive advantage to Mass Mitec at a time when it needed it most and also helped my company to win some prestigious awards for our use of Information Communications Technology. These included an East Midlands Regional Award for entrepreneurial use of Information Communications Technology, winner of the Midland Region Educational section of BTs Broadband Britain award and Runner–up in the National Final of the same competition. The recognition from the national BT broadband competition was especially satisfying and prestigious. The Midland Region Awards Ceremony was held at the Donington Motor Museum near East Midlands Airport and was a highly entertaining affair with many look-alikes of famous movie stars like Charlie Chaplin and Marilyn Munroe and our prize as the winner of the educational section being presented by the former boxer and now television personality Barry McGuigan. Winning the regional event put Mass Mitec forward to be short listed for the finals held at a venue in London near Victoria station. All short listed finalists were provided with hotel accommodation for 2 people from the company and the event was hosted by Paul Ross, brother of Jonathan Ross. There was also great entertainment including some operatic singing from the Four Waiters.

The success of our use of communications technologies for Comknet and our involvement in the use of virtual classrooms for collaboration and training led me to decide that we would focus on Community Informatics as a core business and put all our energies into becoming one of the best specialist consultancies in the world. Comknet had already opened up speaking invitations to events in Brussels and Buenos Aires at the first international Global Community Networks conference in 2000 and even before Comknet had reached the end of its funding period I was already developing a follow on proposal to try to secure finance from a Govt scheme launched at that time and named CALL (Community Access to Lifelong Learning). Using my experience and contacts from Comknet I devised a project called HCLN (Harborough Community Learning Network) which comprised a central learning centre in the town as a hub to a network of rural outposts, an idea which combined the best of Comknet and the South Kilburn Innit project. It was this cunning plan which eventually brought my business to a close and opened a new chapter in my life and involvement with technology.

My initial contact with John Hibbs in the fated global webcast led to a long standing friendship with him and my involvement in one of the most remarkable technology based social enterprise events I have encountered. John is the Founder of the Benjamin Franklin School of Global Education. Just as I had drawn inspiration from Thomas Cook and his passion to do good for his fellow man, John had been inspired by Benjamin Franklin and wanted to do his best to promote the use of communications technology for English Language education on a global basis. He is slightly older than I am and has therefore witnessed even more change than me. He was also strongly wedded to the KISS (Keep It Simple Stupid) and the use of technologies which were as universally available and reliable as possible whereas I have been drawn like a moth to the more bleeding edge and emerging technologies. I certainly shared John's view that telecommunications and collaboration technologies for knowledge sharing should be as inclusive as possible so that as many people could be engaged and have access to any programmes.

John's fantastic initiative was called "Global Learn Day" (GLD). It was run as a 24hr annual event in the October/November timescale and was based on "POTS" technology (Plain Old Telephone System) using a teleconferencing "bridge" to enable as many people as possible to listen and join in to GLD. His inspiring vision for GLD was the concept of a ship sailing round the world to follow the sun and beginning in the Pacific islands. John's energy and passion attracted huge numbers of supporters including some of the most gifted and innovative social entrepreneurs and educational technologists. As well as having access to GLD via your telephone, you could also follow this epic journey visually via your web browser and in many locations listen to it on community radio and television. Over the years John has had many household names as his keynote speakers and he has been able to run this amazing global knowledge sharing exercise with virtually zero budget and infinite goodwill. I had the privilege of John as a guest at our White House home and offices one time and have some treasured memories of his inspiring presence. He is one of the most intellectually generous social entrepreneurs I know and he still retains that spark into his 70th year. Our own efforts at the innovative use of emerging technologies included being one of the first to use streaming video over the internet and I had a webcam permanently set up in our main office to refresh a picture of our main office every 15 seconds on one of our web pages.

The CALL funding scheme had a timeline that would have allowed us to make the Harborough Community Learning Network (HCLN) a natural follow on from Comknet and a continuation of revenue and cash flow. I developed the HCLN bid document and built a partnership that included Harborough District Council and Leicester Mercury and was fronted by the local Council for Voluntary Services (CVS). It was an essential pre-condition of any project that the lead partner had to be a voluntary, public, charity or "not for profit" sector organisation, and not a commercial, albeit small, enterprise like Mass Mitec. HCLN's network hub needed to be located in an accessible and central position in the town

Knoll House on the Harborough Canal Basin – the original planned HQ for HCLN

The Old School House on Coventry Road Market Harborough

and we were very fortunate that a local charitable organisation were willing to use their funds to acquire a suitable building on the site of the canal basin in Market Harborough, not far from the town centre. The property was owned by British Waterways who expressed willingness to lease the building on a very long term basis and negotiations began between the Harborough and Bowden Charities and British Waterways to finalise the contract.

HCLN received its first major setback when British Waterways changed their mind and withdrew the property from the market. My plans had included relocating Mass Mitec to the new building, separating my business from my home, selling the White House and downsizing now that Carol's children were all old enough to leave the nest. This meant a drastic re-think and a frantic search for a new premises and I found a great property even more central than the canal basin. It had been used as the Headquarters of the travel company Travelsphere (to whom I had sold a System/34 computer in my IBM days) before they re-located to a bigger new property on a business park outside Market Harborough. Enthusiastically we approached the Harborough and Bowden Charity with this new proposal but this alternative was rejected by them as it was not such an attractive investment for them. I spent what seemed an interminable amount of time trying to broker a partnership relationship and/or venture capital to acquire the building without success and, over a year after we had our funding bid accepted, our lead partner had to

The Rockingham Road HQ of South Leicestershire College

withdraw from the project because they could not sustain the ongoing costs of trying to sort this problem out.

Once this had happened, Wigston College, which had a small outpost in Market Harborough and was looking to expand within South Leicestershire, agreed to take over the lead role in the project. They had identified a brand new building under construction on the Rockingham Road industrial estate as being a suitable location for their expansion plans. They found funding to employ a consultant, Keith Small, as Project Manager at the end of 2001 and he began to devote his time working alongside me to finalise the contractual arrangements and get the project launched. The contracts were eventually signed in February 2002 with our involvement being through a new Limited Company MM5C Ltd that Keith Small had recommended I set up. The main problem with all these delays and setbacks was that it dramatically affected our cash flow. I had been working almost full time on the project for 18 months without being paid a penny, only made possible by the income I was generating from my HP virtual training contract.

Ironically, it was the 9/11 New York disaster the previous year that proved the final straw and set in motion a chain of events that totally changed the course of my life. One of the repercussions of this terrorist attack was an impact on commercial trading and this was in part a reason why in April 2002 HP advised me that they were cutting back on their virtual training contract with me in order to save costs. It was this cut back in my main commercial income that led to Mass Mitec's removal from the project. It was a tragedy that we were was so close to success in many ways and yet so far in other ways. The new building to house what would become the Market Harborough campus of South Leicestershire College (Wigston College's new brand name) was virtually complete and would be launched, along with the project less than 4 weeks later. I was able, however, as a final contribution to the project, to organise one of my proudest technology innovations "The Radio with Pictures Show" to broadcast HCLN to the rest of the community and to the world beyond.

During the period 2000 to 2002, I had volunteered myself as an ICANN representative. This voluntary organisation's aim was to act as a kind of professional body representing the views of IT professionals in areas like the future internet. It was this involvement that led me to be invited to be part of a European project called "Global Society Dialogue" (GSD) and led to my involvement in Geographical Information Systems (GIS) some years later. GSD was devised and led by the visionary Franz Josef Radermacher of Ulm University in Germany and was intended to raise global awareness of the sustainability challenges of our IT driven society, living in a world with finite energy resources and increasing our consumption levels like never before in history. The project covered all of the travel costs of the dozen or so participants from all over the globe. It was through this project that I met some truly remarkable people who have become lifelong friends. Our first project meeting was in Reisensburg castle in Germany and was a truly memorable affair which brought social entrepreneurs like myself working at grass roots level to discuss and plan strategies to tackle serious social and economic issues.

The workshop concluded with an evening of music played on Mediaeval instruments, a wonderful traditional German meal, much wine and some singing which I initiated by suggesting that a person from each different country represented should sing one of their cultural songs. I began with a loud and probably off key rendition, in true musical hall "Pearly King and Queen" style, of "Maybe it's because I'm a Londoner". This acted as a catalyst for some wonderful and amusing national ditties and some delegate bonding that has lasted many years. Amongst the many unforgettable people at that first workshop were Milan Konecny from the Czech Republic who will feature heavily in a later chapter,

Thomas Schauer, a German working with Professor Radermacher in Ulm, Jyoti Parikh from India, Stella Tabirtsa from Modolva, Milda Hedblom from the USA and Lachse Bahadur from Nepal. I drew heavily on these connections to help me with this innovative HCLN launch event "Radio with Pictures Show".

Undeterred and still relatively unscarred by my previous salutary experience of a global webcast using e-Video, I put together a combination of community Radio, telephone conferencing and virtual classroom technologies to deliver what I still believe is ahead of its time yet a fore-bearer of what will come to pass with public and community sector broadcasting. In my design of the office layout for HCLN, which included relocation of Mass Mitec there and the sale of The White House, I had included a room dedicated to community radio that could be used by our local station Harborough FM who every 12 months were granted a limited licence to broadcast for a limited period to an area which included South Leicestershire and Northamptonshire. Community Radio was an important component in many of the international community informatics projects I had witnessed around the world and I supplemented this with the purchase of a 12 month lease on a HP Virtual Classroom. A large teleconferencing provider called Genesys also supported the project by providing a free teleconferencing bridge for the one week period of the broadcasts.

The Radio with Pictures Show was essentially a breakfast radio chat show that I hosted for seven days between 6am and 7am every morning.

The author broadcasting the Radio with Pictures Show on Harborough FM in 2002

Unlike all the other programming on Harborough FM which was largely focused around music, news and travel reports, my show had only small amounts of music to signify breaks between interviewees. The theme for the week was how Information Communications Technologies were tackling serious social and economic issues in communities around the world with a different topic each morning including ICT for Entrepreneurship, ICT for Education and ICT for Disadvantaged People. Over the 3 weeks leading up to this event I lobbied all my contacts from public, private and social enterprise sectors to submit their ideas and agree to be interviewed on air. The format included two local guests in the studio with me, two UK based experts and two international experts from all over the globe. Each person I interviewed either supplied me with a small number of Powerpoint slides to display in the virtual classroom or approved some content I developed for them. This time the cunning plan worked and the technology worked beautifully with participants as far away as the Arctic Circle joining by satellite radio as well as the more local regular radio listeners.

There were many highlights during that week including a programme on ICT for Entrepreneurship that attracted Debra Amidon, the passionate and highly intelligent founder of a Global Knowledge initiative called Entovation. Debra operates out of Boston Massachusetts in the USA which took its name from my home town in Lincolnshire where some of the early Pilgrim Fathers were imprisoned to prevent them sailing for the new land from nearby Fishtoft Creek in the days long before Boston Docks were built. These founders of the America that became the global superpower eventually made their way by land to Plymouth from where they set sail on board the Mayflower. Boston USA was also home, of course, to the famous Boston Tea Party that signalled the revolution that ended colonial rule from England. With myself and Debra having a common connection with Boston, I also discovered by accident that a third guest on that programme also had Boston connections as she had attended Boston High School for Girls. This happy set of coincidences encouraged me to use the web site Friends Reunited to find ex-schoolmates of mine from Boston Grammar School who were now living abroad. My search turned up a man whose career in the Foreign Office had led him to become the British Ambassador in

Belize. He had been at Boston Grammar during the same time period as myself but was one or two years ahead of me. The latter part of this programme was a live interview with him in Belize and included my showing on the virtual classroom screen images of the Boston United Football team that I had played in and a web page with the words of our Latin school song "Floreat Bostona". He was also logged on to this classroom and the technology worked well enough for us to sing our old school anthem over air to our bemused listeners.

One of the other highlights was in the programme on the theme of ICT and climate change which featured a number of my colleagues from the Global Society Dialogue project. One of these colleagues was Thomas Schauer for whom I had the greatest respect for the way he "Walked the Talk" by living a modest and environmentally committed lifestyle. Thomas defied the common perception held in the UK that the Germans have no sense of humour. The rivalry between England and Germany that is the legacy of two world wars is constantly played out in every single soccer match between the two nations, never more controversially realised than in the 1966 World Cup Final between our two countries. The week of my programme was one of those very rare occasions when the English soccer fan has any reason to gloat. England under Sven Goran Errickson had just beaten Germany 5-1 on their home soil so I could not resist including a slide in the presentation I was sharing during my chat with Thomas that showed a football with some text below that read "Germany 1 England 5. Somewhat provocatively but in jest I asked Thomas if he could see what was on the screen and suggested he might like to read it out. When he appeared very reluctant to share this information with the radio and telephone audience I suggested that it might be because Germans have no sense of humour. His reply was an instant, very sharp and hilarious riposte. "No it is you English who have no sense of humour because when Germany beat an England team managed by Kevin Keegan 1-0 at Wembley you were such bad losers that you tore the stadium down." He was factually correct because I believe that was the last soccer match before the old Wembley Stadium was demolished to prepare the site for the new stadium.

This traumatic period between the end of the Comknet project and the eventual demise of my business did have the compensation that it introduced me to many amazing and largely unsung social entrepreneurs from all over the world. Amongst these individuals was a young man called Simon Stevens who had cerebral palsy and had to spend much of his waking life in a wheelchair wearing a skull cap. Simon is one of the most intelligent and visionary users of Information Communications Technology that I know of. He is a serial entrepreneur by nature and when I was introduced to him by a local disability campaigner, he was already using his PC as a communications hub for almost all the tools available at the time. Because Simon lives in Coventry, I did not meet him in the flesh but had regular exchanges of emails from him which revealed him to be articulate and intelligent. He had long held a passion for encouraging entrepreneurship amongst disabled people and had won a national award presented by Gordon Brown in recognition of his outstanding work. Simon had submitted a set of slides to me for his slot on the chat show and these were so well crafted that I had no hesitation in inviting him to be a guest on the relevant programme. Until this time I had never heard Simon speak so it came as a huge shock to me when I spoke to him for the first time live on air and heard the speech problems that had blighted his life since birth. His cerebral palsy made Simon's words almost unintelligible to anyone listening to him for the first time and had been one of the primary reasons why so many people had dismissed him as having learning difficulties. He told me that as early as the age of three he had been placed in a mental institution and had since then suffered unimaginable discriminations that had been so much of a barrier to him fulfilling his undoubted potential as a human being. Simon was a classic example of an intelligent human using his abilities to make the most of machines. On this occasion, happily, Simon's Powerpoint slides were so well crafted that his speech problem would have been totally irrelevant to anyone logged on to the virtual classroom.

It was the Radio with Pictures Show that also brought me into contact with another remarkable person in the shape of Steve Thompson. I believe that the first time I met Steve was at a gathering in London organised by an international community radio organisation called AMARC. It was clear from our brief conversation at that event that Steve shared the same visions about the use of technology for community development. Something of a maverick, Steve had been responsible for setting up a pirate community radio station in the North East. With a strong Geordie accent, Steve had been born the son of a steelworker at the long-gone Consett steelworks near Middlesborough. Steve's father, as was the tradition in that region, had wanted him to follow in his footsteps and get a job at Consett steel works like so many other families. Steve Thompson had other ideas and his ambitions were more directed to being a rock star. He was able to write as well as play music and he did record one song that made it into the charts in the 1970s. One of his songs was also translated into French and recorded by Celine Dion for one of her albums. This initial contact with Steve was followed by us meeting again at a Global Community Networks conference in Barcelona where Steve high-jacked the grand piano in the five star hotel lobby for some impromptu nostalgic renditions of popular songs by our delegation.

Steve was to feature in my life several times over the decade. He is one of the most innovative and entrepreneurial social entrepreneurs I know. Although this was a very turbulent and in many ways unhappy period in my personal and business life, it did open up many doors and expose me to the wealth of ideas for technology that ordinary people with no technical background can generate. I became involved in a number of funding initiatives around that period, all focused on the use of technology for community development and lifelong learning. One such project was in a European lifelong learning programme called Grundtvig II which was administered in the UK by British Council whose headquarters are in Spring Gardens near Trafalgar Square in London. The project I became a partner in was called "ERDE" (European Rural Development by Education). The abbreviation also means "Earth" in German. ERDE was a so-called mobility project which means that the European Union funding is designed to cover the cost of visiting other European projects for knowledge sharing and dissemination of best practices.

It was ERDE that also extended my network of gifted social entrepreneurs working at grass-roots level. There were partners from Austria, Germany, Poland, Lithuania, Hungary and ourselves from the UK. The Austrian partner contact whom I believe was the driving force behind ERDE was a passionate environmentalist called Franz Nahrada whose family owned and ran a hotel in Vienna. I seem to recall that I first met Franz at a Vienna workshop as part of the Global Society Dialogue project and he was working on his pet initiative called "Global Villages" which was based around the concept of using collaboration and learning technologies to build a more equitable and sustainable world. I think it was Franz who recruited me to be part of ERDE and I was fortunate enough that the project was accepted and, although it did not generate any income for me, it did cover all my travel expenses and created many contacts and opportunities that I still value today.

Each of the partner countries was responsible for organising and hosting a visit from all the other partners to demonstrate examples of best practice in community based lifelong learning. Most of the partner projects revolved around Community Technology Centres like the one I had devised for HCLN but there were different and interesting uses of communications technology in every individual country. Every partner country visit brought with it memorable moments too numerous to mention. All the partners sought to incorporate cultural heritage as a core component of each visit and this merry band of grass roots social entrepreneurs developed mutual admiration and friendship which lasted through the ensuing years. It was the visits to Poland, Lithuania and the UK which hold the strongest memories.

The visit to Poland actually started in Berlin where we all flew in from our various countries and gathered in a main square to board the minibuses that would take us to our university town base for the initial workshops. The co-ordinator for the Polish section was called Waclaw Idziak, an academic who came across as more than a little eccentric because he spent most of his introduction to the group in the first workshop in Vienna teaching us all how to juggle. Games were a very important part of his strategy for ice-breaking and bond-forming and they were very effective. The Poland workshop began in Malechova and went on to visit his project's "Thematic Villages". He shared my belief, embodied in the Comknet project, that every community has its own rich set of talents and that cultural heritage could act as the core of rural development. The Thematic Villages were his "Living Laboratories" that sought to exploit the use of technology to build on community strengths and specialisms.

The first of these villages was using paper making as its core theme.

This rural location had quite a long history of paper making and a legacy of community knowledge and stories build up over centuries. Working with local people he re-invigorated this heritage to get the community behind making paper not only a commercial venture but also a tourist attraction and a skill that was passed down to future generations in the schools. The IT centre set up in the local school was used to research paper and promote the village as a tourist attraction through "Paper Trails" through the local woods. Children were taught the basics of how to make paper from its raw components and took part in practical paper making sessions.

The next village also gave some clue as to Waclaw's propensity to juggle at every opportunity because juggling and circus skill was something this village had been identified with for many years. Working with the local people, Waclaw helped the village to make this something of a differentiator for them that would build a strong brand identity and attract other people to the community with a passion for these arts. Juggling was taught in the schools and we watched displays by the local children. The other thematic village I remember was known as the "Hobbit" village after Lord of the Rings. Many years ago someone had started to build a miniature village based on J.S. Tolkein's book and this became a base from which to build a strong identity and attract tourists to this poor rural region. The Hobbit village was also noteworthy because we had a "Hog Roast" in the evening with a whole wild boar being cooked on a rotating spit. The Poland trip was completed by a discussion and networking workshop in the North of Poland at a tiny seaside resort.

Lithuania was the other country that held many pleasant surprises. Although I had never visited the country before, I had a firm impression in my mind of a grey communist bloc country with functional buildings. I arrived in Vilnius after an inexpensive flight to be set back by how charming and historical this city is. The airport must be amongst the closest to any capital city I know (apart from San Diego where you appear to fly between skyscrapers). I stayed in a small B&B on Pilimo street near the main station and although this area was somewhat run down, it was short walking distance into the Old Town and charming squares and narrow streets, totally unlike the picture I had painted for myself. It was a very short stay in Vilnius before I caught the train to the rural villages which would be the location of our stay. This journey from Vilnius's main station was unlike any I had ever encountered, even on the pre-Beeching rural branch lines with their little village halts. The train ambled casually on its way into forest countryside where it periodically and regularly stopped in the middle of nowhere to set down and pick up people who were berry picking in the woods. At these spots there were no platforms or station signs to indicate civilisation although the train did stop at some village stations.

Our hosts had arranged for us to stay in little Alpine chalets in the forest, complete with a small communal open air swimming pool and, yet another surprise, blue skies and pleasant warm weather in late summer. As well as the grey communist building legacy, I had expected cold weather and leaden grey skies from a country further north than my own. During this short visit we saw many projects that were linked to cultural heritage and the use of ICT to develop social and economic wealth in areas where traditional industries and crafts were in decline. One of these highlights was a theatre performance which took place in a building much like a huge barn in a forest clearing.

Eventually it came to my turn to host a visit from all the other partners who had all, in their various ways, proved to be a hard act to follow. Pride and hurt at being sidelined from the HCLN project that had been "my baby" from the beginning prevented me from hosting the visit in my own region so I turned to my Geordie friend Steve Thompson. By this time Steve had moved from his freelancing activities to the more secure paid position of a Community Informatics Project Manager at the University of Teesside in Middlesborough which had its own and very unusual Community Informatics Department. Steve was more than willing to help and I used the UKs share of the project budget to help him organise a series of showcase activities in the North Yorkshire and Teesside area. Steve's job at the university required him to set up community technology centres in some of the communities hardest hit by the decline in their traditional mining and steel industries. He acted as a grass roots evangelist who, like me, sought out community champions to train to make these hubs attractive to local people and sustainable once the funding dried up.

Steve was perfectly suited to this role with his down to earth approach and creative talents. Through the university admin I had arranged accommodation to fit the modest budgets of the partner countries and also hired a minibus to collect the delegates from Teesside and Newcastle airports as well as ferry them around on the visits Steve and I had arranged between us. It was quite late in the evening when I pulled the minibus into the small driveway in front of the Bed and Breakfast I had pre-booked. There was uniform dismay from everyone when they saw a bucket sat on the dining table to catch the rain that had been leaking through the roof. This, added to the very basic nature of the rooms and Franz's first-hand experience of the hotel business, almost caused mayhem but to everyone's credit, including the hotel's, immediate issues were resolved and the traditional full English breakfast the following morning made the world seem a far better place.

Steve had arranged visits to the former mining communities where, as in all the other countries' projects, cultural heritage played a large part. The local people made everyone welcome and one of the early highlights of the visit involved an example of Steve's creative use of new media and telecommunications technologies. Steve had organised a "Virtual Pub Quiz" to take place online in several pubs around Middlesborough and the more remote areas. In each pub there was a computer connected to the internet, a data projector and a screen. The partners from the different countries were split up and joined residents taking part in those pubs. Everyone in the pub could join in answering time limited questions which were submitted over the internet and collated by Steve's team. It was a fantastic night and a superb example of the unifying and regenerating potential of community informatics.

For those delegates from other European countries, the highlight was probably my piece of self-indulgence in taking them to some of my favourite spots in North Yorkshire. We visited the quaint market town of Richmond which they all loved for its quintessential Englishness and then back across the moors to Goathland, home of the Television series "Heartbeat" with the Aidensfield Arms pub, iconic Post Office and Scripps's garage, all very well known to millions of viewers in the UK. I took them for an English tea to the Mallyan Spout Hotel where many of the Heartbeat cast used to stay during filming but then came the most photographed visit of all – Goathland Station on the North Yorkshire

Moors Preserved Steam Railway which doubles not only as Aidensfield station in Heartbeat but also Hogwarts station in the Harry Potter films. I was able to tell them some of the stories about the filming of Harry Potter that my SFX specialist friend Bob Bridges had passed on to me.

I had reason to visit Steve at the University of Teesside on another occasion connected with his work on Community Informatics when he had organised a project Open Day on the campus there. One of the modern buildings at the University was hosting the event which had lots of stands manned by the community projects he had helped implement. For me, this day was a magnificent example of the empowering nature of technology. This most modern of universities, set in its heartland and proud industrial heritage, had opened its doors to the citizens from all walks of life many of whom, like my father, had dreamed of being able to send their children to university to give them a start in life that their backgrounds and lack of opportunities had denied them. All of the villages had stories to tell, not only of the history of the community, but of their own lives, of what life was like for them before and during the war years when the mines and steelworks were flourishing. There was an unmistakeable pride in their voices that they felt worthy at last to be standing alongside academics and professors. Steve also showcased some of the projects with keynote speakers in the auditorium, including a live video link with Howard Rheingold from the USA. It also included, inevitably, some of Steve's songs about his beloved region. His life has been so rich and full that I often joked with him that he should write a musical about it – which he later did.

The ERDE project was not only a great success for all concerned, it was highly enjoyable and this period was probably the start of the regular international travel I have been lucky enough to make over recent years. In the February of 2002, ironically just before the problems, I was invited to speak at an international conference on regional ICT projects in South Korea. This was the first of what became many visits in the latter half of the decade. I was flown to Incheon airport near Seoul to be met at the airport by some students who had driven up to collect myself and another presenter who was flying in from India. Once his plane arrived we set off for Kyunju, the location for this conference.

Kyunju had been the ancient capital of Korea and was over 100 kilometres away. The route to Kyunju took us through rush-hour Seoul so the students suggested that we ought to stop and have a meal whilst the traffic died down. This was in 2002 when the Soccer World Cup was being jointly hosted by South Korea and Japan. Everywhere there was evidence of feverish construction activities to create the roads and stadiums for the tournament. The restaurant served traditional Korean food but had tables and chairs unlike the really traditional Korean restaurants where the table is very low and guests sit cross-legged on the floor. In the centre of our table was a large circular hole and I began to recall one of Michael Palin's travel documentaries which featured stories about monkey brain being a delicacy in some parts of the world and involved slicing the skull and eating the still warm brain like a soft boiled egg. It was a great relief when this circular hole was used instead to house a small charcoal grill which we would use to cook our own meat on.

We finished the latter part of our journey by bus and eventually arrived at the conference venue, a hotel complex beside a very pleasant lake. My presentation featured my work on Comknet and the Harborough Community Learning Network and drew heavily on the Thomas Cook's story as a model of how communications technology could transform society and help build social and economic wealth. South Korea was and still is an acknowledged leader in the use of telecommunications as an engine for economic growth but this conference gave me an understanding of how they had transformed their country from being a poor nation recovering from a massive civil war into an economic powerhouse. The Saemoul movement was credited for this success and involved giving citizens the vision and motivation to create the essential infrastructure with roads and railways and telephone networks. The conference I was attending reflected a rebirth of this movement but using the internet and community web portals developed by citizens in much the same way as I was promoting both ComKnet and HCLN.

My work on Community Informatics within my local area over the period 1998 to 2002 was drawing to a sad conclusion. It had opened my eyes to the empowering nature of technology placed in the hands of ordinary citizens and it had gained me entry to a world of international travel and conference speaking. The following years from mid-2002 to 2005 became a battle for survival and one of the few periods in my life that give me cause for regret. I had established a strong reputation seemingly everywhere in the world except my own back yard and the regret I feel is for the effort I wasted in trying to get compensation for all the time, energy and passion that I had invested in conceiving and developing the Harborough Community Learning Network.

I was helped by many people during this period. I am grateful to Ian Vickerage for his company's loan of £10k, to James Briggs of Harborough Rubber for his generous commission for my work, to Caryll Stephen of FWR, my most valued client, and to Venu Dhupa who had been Chief Executive of Nottingham Playhouse and recruited me to be a regional champion on an East Midlands Development Board before later giving me a chance to present my Radio with Pictures Show idea for a NESTA funding award.

It was in 2003 that I really began the next chapter of my life and had my awakening to the changes in power that this book reflects on. It was during this time that I had my first experiences of the intelligence of machines being deployed to get the best from "less intelligent" human beings. I was now working on my own trying to earn enough money to pay the mortgage and bills. I had already applied for a number of jobs and had been shortlisted to positions which included a consultancy post in the Deputy Prime Minister's office and a similar post at the British Council in London. I came close in all of these but not close enough. In desperation I responded to an advert in the local paper for night shift workers in a Daventry Frozen Food Warehouse. The job involved 12 hour shifts from 7pm to 7am for four nights on and then four nights off. This seemed the perfect answer to my problem and would allow me to bring home a regular and guaranteed monthly salary from night work and still leave me free time to do consultancy work to fill the void. The detailed story of my time at this warehouse is included in the next chapter on computing technology. The experience of working in this Tesco warehouse became a valuable part of my lessons in life and served me well for what was to come next.

During the two years I was doing night shift, I was continuing to earn income for my virtual classroom field training work with Agilent Technologies and also supporting my main commercial client, the Foundation for Water Research (FWR) whose Chief Executive Caryll

Stephen shared my enthusiasm for the strategic use of telecommunications technologies. Caryll and her husband Derek lived in a converted chapel in the village of Moreton near Thame in Oxfordshire. FWR's main office was about half an hour's drive away in Marlow, Buckinghamshire along the M40. Caryll commissioned me to set up her home office and support a range of different technologies both there and in the main office in Marlow. Her story, relating the various technologies in this book, appears across many of the chapters as she followed me down the bleeding edge path, ever seeking to stay in step with whatever could help her and her organisation to exploit their full potential.

Caryll appreciated the importance of telecommunications within FWR, not only to help in the dissemination of FWR's research activities, but also as a tool that allowed her to conduct her business from anywhere in the world. In the days before Bluetooth wireless technology became available, it was possible to wirelessly connect devices by using Infrared. Caryll bought a tiny Toshiba notepad computer from me in the 1990s which she used to connect to her mobile phone via infra-red and connect to the internet using the phone as a modem. One of her proudest achievements was the time she was able to send digital photos to her office using this technique whilst she was in a very remote forest location. Today, at her new home near Malmesbury, her mobile signal is so poor at times that this sometimes proves impossible in the UK – so much for progress!

It was in the summer of 2005 that I saw the job advert in the Leicester Mercury that set me on the road to recovery in both my business and personal life. De Montfort University in the centre of Leicester had secured funding to set up and operate a Creative Industries Knowledge Network, an initiative to try to build bridges between industry and academia and boost the capabilities of both sectors by making research knowledge accessible to small creative businesses and exposing students and academics to the realities of commercial life. In the time since I had been at University in Birmingham and later studying the Diploma in Management studies whilst working for Post Office Telecommunications, successive Governments had sought to improve the nation's wealth by making higher education available to as many young people as possible.

Leicester Polytechnic, like many other further education colleges around the UK, including Lanchester Polytechnic in Coventry, achieved university status and had been renamed De Montfort University. By this time my first wife Angela had long been in post as Director of Physical Education at Leicester University and now I was about to re-enter academic life at her rival university. The advertised job was a two year project management contract to set up this Creative Industries Network and run events and activities that shared and developed knowledge across both sectors. I went for the interview in the Fletcher Building in the heart of the campus more in hope than expectation. At the age of 56, I had seen my job prospects receding and had already missed out on another job application at nearby Leicester College. I was interviewed by Iona Cruickshank and Gerard Moran of the Arts faculty and as I drove back to Lubenham I was praying that I would be successful this time. The university contacted me the same day to offer me the job and within a few weeks I was occupying a new small office on the 7th floor of the Fletcher building.

After all my experience of Comknet and HCLN and the physical demands of the frozen food warehouse, my new job seemed as easy as shelling peas. I had some fantastic support from Iona, who was my immediate boss, and her team of lecturers. Iona's specialism was photography and her passion for the subject is covered in more detail in the Imaging section. In taking on this job I was determined to use technology as creatively and cost effectively as possible to use public money to help small creative businesses, such as my own had been, to prosper, and one of my first uses of my budget was the acquisition of web conferencing technology. I chose Adobe Connect Professional as the platform of choice for a number of different reasons. Apart from the fact that Adobe were very supportive and pro-active in promoting the use of their creative products both in universities and industry, Connect Professional has some very useful and powerful features that I wanted to exploit. Amongst these features were tools that allow the management of email invitations and registrations to webinar events and the facility to use Microsoft Powerpoint to create and record presentations with synchronised audio and video. I used this pre-recording tool to set up a mini promotional video about this Creative Industries project that could be accessed and viewed through the Connect Professional virtual classroom.

It was in this job that I also renewed my relationship with two people I had connected to directly and indirectly as a result of "The Radio with Pictures Show". The chat show that featured my home town and Boston USA, and was discussing entrepreneurship, was the beginning of an ongoing relationship with Debra Amidon, the founder of the Entovation Global Knowledge Network. Debra had honoured me by appointing me as one of her elite e100 knowledge professionals on the Entovation web site and had managed to get funding to enable me to travel to the First Global Entrepreneurship Summit in Muscat, Oman. This incredible event brought entrepreneurs and those working to promote entrepreneurship across all sectors from all over the world to this exclusive beach resort a few miles from Muscat City. It was at this conference that I had the privilege of joining a breakfast seminar on the beach given by the famous visionary, Deepak Chopra, who was talking about his theories on parallel universe and the concurrence of the physical and invisible virtual energy based worlds that accounted for simultaneous inventions throughout history with no apparent connection between them. It was his reflection on the concept of synchronicity and coincidence that made a big impression on me because I have always felt that I have been a very lucky recipient of amazing coincidences, like finding yourself next to someone on a train who turns out to have unexpected and statistically unlikely connections to your world. He said that this phenomenon actually happens to everyone but only people who are open to such possibilities are able to recognise and exploit them.

The Oman conference introduced me, through Debra Amidon, to the second person in the shape of David Pender, an "ex-pat" UK citizen who had worked in Oman for a number of years. Meeting David Pender was another classic example of synchronicity that Deepak Chopra referred to. It was literally on the last day on the Oman conference in the main hall as everyone was milling around saying their goodbyes that I found myself standing next to what looked like a fellow Englishman. In the brief conversation that followed, Dave explained that he and his wife and two sons had lived in Oman for a number of years and his job involved helping the Govt develop the industries and economy that would sustain them in the post-oil world. He was responsible for attracting new technology based business to the region and helping to develop a global brand for Oman as a major player in these new economies. We got on very well and he saw the relevance of my work with Community Informatics to the challenges his adopted country faced – a cash rich nation with a high population of young people for whom the traditional Government jobs would not be available when they finished their education. This brief contact with Dave proved the start of a long relationship with Oman and a collaborative agreement to jointly develop their Knowledge Oasis Muscat (KOM) project after I became Director of the Serious Games Institute.

It was whilst I was at De Montfort University (DMU) that I came to work with Michael Powell who ran the Game Art Design courses there and had students available for placement in creative industries. In my eye, Michael bore a striking resemblance to Michael Palin, the Monty Python star. Michael had developed strong connections with Codemasters, the Midlands based global games publisher and it was through Michael that I had my first real exposure to Blitz Games who were to become a very important part of my work at the Serious Games Institute later on. Blitz Games, whose headquarters is in Leamington Spa, held annual Open Days at their offices for both academics and students from Higher Education. The aim of these events was to build better understanding and collaboration between the games industry and higher education and was a highly organised a professional insight into how one of the largest and most successful independent games companies operates to deliver the video games that entertain millions across the globe.

The other important contribution Michael made to shape my future activities was to introduce me to the virtual world "Second Life", developed and managed by Linden Labs. Michael sent me a link to their web site, which I used to create my own avatar or alter ego, Hobson Hoggard. The choice of this combination was based on the availability of the surname of Matthew Hoggard, the England bowler who had helped our cricket team win the Ashes test against Australia, and the fact that I had recently played the part of Hobson in the stage version of "Hobson's Choice". I used this introduction to Second Life to explore something of the potential of virtual world technology but it wasn't until I joined the

Serious Games Institute that I began to make practical use of its communications and collaboration capabilities in my job. It was this introduction to Second Life and sharing this information in one of my regular monthly e-newsletters that also helped to change the life of Simon Stevens, the Coventry based entrepreneur with cerebral palsy.

I ran fairly regular seminars in my time at DMU, inviting creative industry speakers and academics to share the latest developments in new media technologies. DMU had its own virtual reality "cave" which it used in a 3D fly-through of proposed developments in Leicester around the main railway station. I have to say that watching this helicopter view flying and banking around the city made me feel slightly ill, and my eyes and my body's sense of balance came into conflict. The Creative Industries Network project introduced me to many talented and interesting people from both outside and within the university. Martin Richardson showed me his highly innovative 3D imaging technology that could be used for both art and commercial purposes such as detecting fake fashions through the use of special 3D embossing on the fabric threads. DMU can also arguably call themselves the pioneers of 3D television without the use of special glasses.

It was in the Spring of 2006 that I organised a "Serious Games" seminar for DMU and invited guest speakers who would also figure in my life in the years to follow. Kam Star of the London based Playgen and Martine Parry, who set up the Apply Group and is an active evangelist for immersive technologies, both made presentations for me and I also organised a remote presentation using Adobe Connect from a games company in India who had developed games based learning programmes for some education courses. I didn't know at the time that a few short months later I would become more deeply involved in serious games technologies myself. It was actually my boss Iona Cruickshank who had spotted the job advert for a Director to set up a brand new Serious Games Institute at Coventry University, and it was she who asked me if I had seen the advert. I am not sure whether she might have thought that I had organised my serious games seminar to help me get this new post but the reality was that, if she had not told me, I would never have seen the details of the post and applied for it.

I was at a conference in Auckland New Zealand when the deadline for applications for the Director post was due. The job was a considerably higher salary than I was earning at DMU and fitted my background and experience perfectly but I it was on a spur of the moment decision that I put my application together from my hotel room in Auckland without any real expectation that I would get the job, especially as I approached the age of sixty. The story of how I came to get the job and what happened whilst I was setting up the SGI is more fully covered in the games technology chapter. It was inevitable though that the telecommunications technologies that had formed the starting point and bedrock of my career would also feature in the development of the Serious Games Institute. I was responsible for choosing the technology infrastructure of this new venture and, as I had done with the White House offices almost twenty years before, I tried to put in place facilities which could grow and be as future-proof as possible. This included a Cisco switching system and extensive Category 6 ethernet cabling and sockets to handle high speed speech, data and video, a seminar room equipped with a webinar control desk to support hybrid seminars and workshops with participants over the globe, and a high tech boardroom equipped with the latest Cisco "Telepresence" technology designed to recreate the physical experience of face to face meetings between remote participants.

Over fifty years have passed since my first abortive attempt to use a telephone box to pass on an important message to my parents using an analogue line carried over copper wires on a point to point connection from the red phone box with its Push Button A to my house just two miles away. It is over twenty five years since I set up my Mass Mitec business and used my desktop PC to send telex mailshots to clunky printers, and fifteen years since I first connected to the internet. In all this time, we have moved from copper wire in the ground to fibre optic cables and wireless networks able to connect not just our voices but our video and our data to many people simultaneously. No longer do the electronic signals that we create when we connect to other people across the globe travel over a single connection point to point but are split up into packets of data that take myriad simultaneous routes to the final destination. No longer do I have to shout into a Bakelite handset to make myself heard at the other end of a crackly line or type text commands to programme a modem to make a data connection for me. I am now able to use plug and play technology on my computer or mobile phone to have a video conversation with my loved ones from almost anywhere in the world.

Telecommunications technology has come a long way over these years in mankind's never ending quest to make technology go faster, be more powerful and have greater capacity. I have seen remarkable changes in the ability of these technologies to carry richer and better quality and now digitised information to its global users. Despite all these advances and those we can envisage in the future, I don't believe that it is telecommunications technology which now provides businesses with competitive advantage or will be responsible for some of the challenges we human beings are likely to face over the coming years. All these opportunities and threats are being created by the emerging technologies that form the basis of the following chapters.

CHAPTER FOUR
Leapfrogging with Ghengis Khan on Level Five

Those of my generation will probably have noticed a musical theme to all of the previous chapters with by-lines taken from words of songs by Pink Floyd, The Who and the Bee Gees, all of whom I am old enough and lucky enough to have seen live in their heyday. This chapter brings the first exception to this train of thought and anyone who does not know me well is unlikely to have the faintest idea why I chose "Leapfrogging with Genghis Khan on Level Five" as the sub-text for this chapter. The reasons have nothing to do with music and are twofold. The first reason is that many friends have encouraged me to write my autobiography and when asked what I would call the book, my answer would be "Leapfrogging with Genghis Khan on Level Five" and then I would go on to explain why. The second reason is more profound and relates directly to a salutary chapter in my life when my career suddenly switched from using my intelligence and knowledge to get the best from machines to a totally different job in which the intelligence and knowledge of machines was directed at getting the best out of me. That part of my journey is revealed towards the end of this chapter so read on!

I was born in 1949 – the same year in which what was possibly the very first electronic computer ever trialled in the UK. Principles of computers had been understood earlier than this and incorporated into machines like the one used at Bletchley Park in the war to crack the German Enigma encryption code. The Ferranti Mark I at Manchester University required a team of human beings to get this early computer to perform even simple mathematical tasks. It wasn't until I went to university that I even came in remotely close contact with any kind of computer. Before that time, my only encounter with anything that might resemble a machine that could compute was when I was courting my very first serious

Ferranti Mark I computer at Manchester University in 1949

girl friend Jenny (who later became the mother of my only child in 1967) who worked as what was known as a Comptometer operator.

Part of my Electronic and Electrical Engineering Degree Course at the University of Birmingham required me to learn some basic computer programming in a language called Fortran. All students on my course were required to write a program to the specification given by our tutor and do what was necessary to get it working properly. Unlike today, programs and data were input into these early computers using what were known as "punched cards", each of which was created by an operator typing our programs into a terminal which then created rectangular thin cards with holes punched them in sequences that the computer could understand. The punched card principle bears an uncanny resemblance to the ancient "hurdy gurdy" pianos which would play tunes controlled by punched cards that were fed into the piano on a spiked cylinder driven by someone turning a handle on the side of the piano.

Once you had written out your first effort in longhand on paper, it was sent to the "computer department" for punching in and then a week later you got back a pack of the punched cards and a print out on big sheets of paper saying "fails to compile" (the computer's way of saying this is a load of rubbish that I don't understand) and showing where your program had errors in it. You then had to correct the errors, once again handwritten on paper, before it was sent back once more for retyping and this iterative process lasted until you finally got your program to successfully complete its task. All this happened, incredibly, only forty years ago. The other exposure to computing during my time at Birmingham was in my final year project when I had to build an analogue computer. This sounds today like a contradiction in terms as most computers today operate entirely digitally. My device was basically a closed loop feedback control system of the kind which might be used to control steering in motors.

These were definitely the days when it was the intelligence and skill of a human being that could get the most out of the machine that was the computer. There were only a few primitive computer languages in those days of which Fortran would be described as a "higher level" language – which effectively meant that programs could be written by "lower level" people who didn't know or want to get involved in the complexities of fully understanding every nuance of what happened inside the box. The computer geeks or "petrol heads" of those days would have understood this and would "code" in the Assembler Language, which was the closest human beings could get to talking the actual language of the machine. This was a real and very important skill at that time because early computers did not have the power and capacity of modern systems and had to be nurtured very carefully to be able to handle more complex or heavy duty tasks.

After I left University and started work in Leicester for Post Office Telecommunications, I vaguely recall being sent on a computer basics course designed to give me a technical understanding of what went on inside a computer's memory but this technology, as I neared the end of the 1970s, was still largely the domain of big companies. Despite this, I did recognise in those days that there would be a convergence of telecommunications and computing technologies and that this could lead to business opportunities. I was becoming bored and frustrated at my life as a Management Tutor down in Cooden Beach near Bexhill on what was known locally as the Costa Geriatrica because of its popularity as a base for retirement to that "Dumroamin – dream home by the Sea". Even the delights of Bexhill's Indian restaurant, with its patterned red flocked wallpaper, were no longer sufficient to keep me there. My introduction to the mouth watering passion for curry had only been a few years earlier through one of my staff, Ian McGowan, a "died in the wool", sweat pouring off your head, vindaloo curry aficionado. Even today I rarely venture beyond the mild Chicken Korma or Biryani heat plateau but they are a weakness of mine and I am a regular at the Dhaka Deli on Narborough Road in Leicester.

It was the combination of career frustration and boredom with my job that led me to apply to be a computer salesman, and so I had interviews with NCR and IBM. My interview with IBM was in Nottingham, the branch where this vacancy existed. Looking back, my attire for the interview, although a suit and tie, was unfashionable and ill-suited to a job which involved selling to senior executives. My interviewer, who would become my immediate boss, kept calling me "young man", perhaps through the influence of Nottingham's footballing deity, Brian Clough. The weakness in my application for the Sales and Marketing Executive post was that I had only worked for a public sector organisation and, although I had a distinction in my Diploma in Management Studies and had acquired membership of the British Institute of Management, the practical running of a private sector business was outside the scope of my experience. Fortunately for me, and one of the reasons why I chose IBM, my new employer had a justified reputation of offering the best training in the industry.

Once I was offered the job with a reasonable basic salary and good "OTE" (On Target Earnings), I had the task of telling my Post Office Telecomms employees that I was leaving and explaining the reasons why. This lead to a "leaving interview" with someone two ranks higher than me (equivalent to a Deputy General Manager in a Telephone Area). He had my latest job appraisal in front of him which had excellent rankings in almost every aspect but still showed "Not yet ready for promotion" at the bottom of the form. When I told him that I was ambitious and wanted to go as far and as quickly as my abilities would take me in my career, his reply was quite short and to the point. He said "The Post Office has invested a lot of money in training you and has given you two scholarships. I need to understand the reasons why you want to leave so that we don't employ someone like you again." On that sad note I packed my things and set off back to my home in Great Glen, Leicestershire, ready to set out on the next part of my journey.

My IBM office base was on Maid Marian Way in Nottingham city centre. I was employed by IBM General Systems Division (GSD) which focused on mini computers and mid range systems. Our big brother mainframe Data Processing Division (DPD) was only two hundred yards down the road at City Gate. IBM's other division at the time was Office Products Division (OPD) and sold mainly office equipment like golf ball typewriters and early word processors such as "Displaywriter". Like the famous television sketch which had John Cleese, Ronnie Barker and Ronnie Corbett in a line of clearly diminishing heights and state of dress, DPD looked down on GSD and GSD looked down on OPD. Within the GSD Nottingham office there was a similar kind of hierarchy in which the higher paid Sales Executives looked down on their more technical Systems Engineers (SEs) whose principal role seemed to be stopping the less technical Sales and Marketing Executives using their silver tongue to sell technically impossible or impractical solutions.

Appointed at the same time as me was another trainee Sales and Marketing Executive called Doug Hall, a ginger-whiskered character who, unlike me, did have previous sales experience albeit in a different industry. I was to share my sales training journey with Doug as together we attended the many training courses for both selling and technical skills that IBM offered before putting you out into the real world "on quota". As soon as I joined my team in Nottingham I was paired with an experienced salesman called Cliff Alsop as his "leg man" – to not only learn the ropes from him but also do all the less interesting support work required in closing deals. Like me, Cliff had connections with Lincolnshire where he lived in North Hykeham, between Sleaford and Lincoln. His sales patch was also in this part of the world, extending north to the Humber. I was to learn a great deal from Cliff who was a successful, professional and hard working guy who, like many in IBM, had a dry sense of humour.

It was very early on in this new career, almost certainly within the first two weeks of my time in Maid Marian Way, that I had a couple of experiences that stay with me all my life. The first occasion was when Cliff asked me to "baby-sit" a potential new client who had been invited to one of IBM's sales events in our offices to hear about the new IBM System/34 which I was to learn would be the bread and butter of most

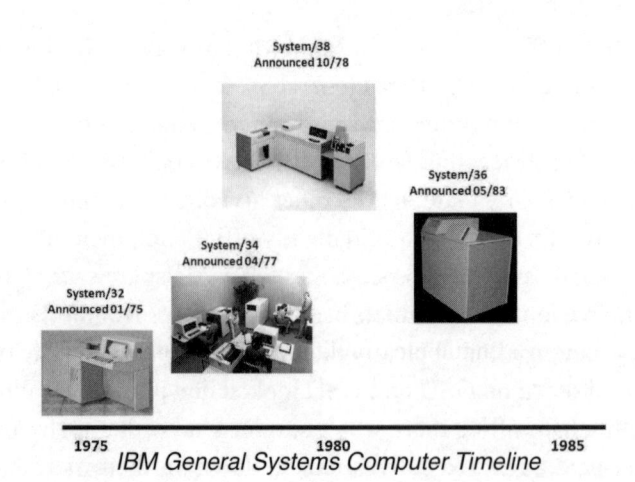

IBM General Systems Computer Timeline
(System/32 Announced 01/75; System/34 Announced 04/77; System/38 Announced 10/78; System/36 Announced 05/83)

GSD sales activities over the coming years. The System/34 or S/34 was the first multi-terminal mini computer IBM had launched to address the potential of computers for medium sized businesses who could not afford or warrant expensive data processing staff (now known as IT professionals) or special air conditioned rooms to house them. Unlike their big brother mainframes such as the IBM 370 sold by DPD in City Gate, the S/34 was quite happy in a normal office environment and was not much bigger than a small row of table height filing cabinets. The S/34 was the much heralded replacement for the single user IBM S/32 which had enjoyed considerable success in the same market over previous years.

Even though I was very naive and inexperienced, Cliff felt that I could do no damage to his prospects with this potential client at such an early stage in the sales cycle, especially when most of the selling would be done by the presenters at the seminar. My job was simply to make the visitor feel important and that his potential business was valued. I thought it very little out of the ordinary when the potential client said he would like to order a System/34 at the end of the seminar. When I told Cliff the next day and he nearly fell of his chair in surprise, I realised that selling was not going to be as easy as this "bluebird" (the name we gave to unexpected and rapid sales to previously unknown potential customers). The following week we both went to visit the company, Wold Farm

Foods, a frozen food manufacturer and distributor to discuss the specifics of the order and get the contract prepared for signature. This was the first of many fantastically educational and interesting visits to dozens of medium and large size businesses that I was destined to make over the coming years.

I was totally absorbed in their description of the processes involved in getting the vegetables from the fields to their freshly frozen state in our shops. I listened in awe at this previously unknown to me and unconsidered by me challenge of retaining maximum freshness. We were shown a measuring device called a "tenderometer" which did exactly what its name implied. It was used by the farmers in the fields to test the tenderness of the peas to gauge the exact moment when the peas were at their most tender, day or night, at which point the massive harvesters that had replaced the manual labour activities, which had long since earned me my "Airfix" Wellington bomber kit from pulling peas manually, sprang into action and loaded wagons for immediate transportation to Wold Farms factory where they were frozen and packed within 24 hours of that "sweet spot" of tenderness. As I recall it, Wold Farms bought their IBM S/34 initially, like so many clients in those days, to handle the accounts processes including sales, purchase and nominal ledger. The specialities of their business would require special application development by the small software houses that acted as IBM partners in those days.

The second experience taught me huge amounts about selling and about technology demonstrations, some of which served me well in later years, and some of which I ignored at my peril on other occasions. I was invited to sit in on a demonstration of an IBM System/32 by a very experienced and naturally charming salesman called John Newbery, known very unkindly and inappropriately (apart from his height challenge) as "Small and Evil". John had been with IBM several years and had "graduated" from Office Products, where he enjoyed years of success selling the iconic IBM "golf-ball" typewriters, to GSD. The demonstration he had arranged at a meeting which would "close the deal" was to a footware company called "Towles" whom I believe were based in Loughborough. The Directors of this company were personal and social friends of John's and he had persuaded them that the IBM S/32

IBM System/32 computer – 32k memory 27Mb Winchester disk

was the perfect computer to handle not just the basic accounting functions but also the more complex manufacturing processes such as stock control and material requirements planning for which IBM offered an "off the shelf" package for the S/32 called MMAS.

To understand, especially in these days of gigabyte memories and terabyte disk drives that are sold in stores like Aldi, the significance of what was being sold in this Nottingham office in 1979, I need to describe the S/32. It was the size and height of an average dining table. It was intended bring business computing to the masses by virtue of its lack of a requirement for air conditioned rooms such as the mainframes had required. It incorporated a "Winchester" disk drive in a sealed plastic bubble and this and its main memory were contained in the side cabinet whose door could be easily opened to reveal the miracle of this hermetically sealed long playing record sized disk spinning impressively with (when running an application) its access arm darting backwards and forwards across the surface of this disk as it "Random Accessed" data like a manic stylus on my record player at home. The maximum configuration of the S/32 computer was 32 Kilobytes of main memory and 27 Megabytes of disk memory and this was sold at around £27,000 with its applications at the time (or could be leased on a monthly basis). During the following two hours of this "demonstration", John showed off the impressive interior of the machine and the compact "Star Trek" like monochrome text display mounted onto the desk and described the benefits of this computer to his potential client friends. At the end of the session he was able to complete the formalities of the contract and it was not until long after this success that I realised the secret of his highly personal demonstration style – he didn't switch the S/32 on!! If he had switched the machine on he would have been exposed to the twin embarrassment of the slow response times and his lack of in-depth technical knowledge which normally would have been provided by an accompanying and watchful Systems Engineer.

IBM provided some great training courses and a legion of anecdotal stories that would have made a good "Confessions of an IBM Salesman" movie at the time. There was some excellent sales training provided by the Huthwaite Research Group whose analysis of what makes a successful sales call was developed into a very effective process called "SPIN" which stands for Situation, Problem, Implication, Need and represents the typical pattern of good sales cycle. The objective of SPIN is to get from the Situation stage to an identified Need as quickly as possible and good salesmen could not only find the fastest route but also assess where the potential customer is during the sales cycle. A lot of the sales training at IBM involved dummy sales calls based on actual case studies of client accounts. There were many pitfalls in maximising the effectiveness of SPIN amongst which was not recognising when the customer was ready to sign on the dotted line and so spending so much time understanding the client situation and asking questions that the customer got fed up and changed his mind. At the other end of the SPIN scale was making unjustified assumptions that the customer had a need (because every other company like his did) and not spending time asking about his situation, teasing out problems, turning these into implications for his business before justifying his specific need for the IBM solution.

Doug and I spent many sales and technical courses together where our assessment on dummy sales calls in which the tutor played the role of the potential customer was made on a scale of 1 to 10 based on how much the potential client's impression of IBM had improved (or worsened) by your call. At the bottom end of the scale was the rating called UDH (Unsatisfactory Did Harm) where your contact with this business actually damaged IBM's reputation. These dummy sales calls were normally observed by the two other trainee salesmen in the batch of three calls, each of which involved a different case study. I was present when the tutor gave a UDH rating to a salesman from Bristol called Jim Galley, who was a sufficiently good cricketer to play for Somerset County Second team. Jim's dummy call did not go down well and resulted in a heated exchange between him and the tutor that almost resulted in blows.

Doug was a motor enthusiast and a bit of a "petrol head". I remember him telling me about one incident when, prompted by my recollection of my lanky scoutmaster's Messerschmidt bubble car, he told me about an accident he had in his bubble car when he took a corner too quickly and ended up overturning the car leaving it standing upside down in the road with its wheels spinning like an upturned beetle unable to right itself. I can imagine his embarrassment at having a team of passers-by turn his car over so that he could extricate himself. It was on an IBM presentation skills course at Sudbury Towers in West London that his love of cars led to a very amusing (but not for Doug) incident. It was his turn to do a presentation to his course mates in a room in one of the top floors of this tower block. He had just begun to speak when he glanced sideways out of the window towards the car park below, did a rapid comic "double-take" at what he saw, shouted "Shit" and ran out of the room towards the lift, leaving the tutor and the rest of us somewhat bemused. It turned out, when he returned almost tearfully back to the room, that he had seen his beloved Ford Capri being driven away by a thief and had desperately and unsuccessfully tried to intercept the getaway. The theft was reported to the police and happily his car was recovered undamaged some miles away but unhappily minus his treasured rock music cassette tape collection. I was to suffer a similar experience in reverse some years later in Market Harborough when my stolen car was recovered undamaged but with one of the thief's tapes to supplement my own collection.

I had many wonderful experiences during this training period which also saw my very first computer sale acting solo. I think I was still shadowing Cliff Alsop at the time because the company in question was in his patch at Skegness. It was a tiny enterprise called R G Mitchell which also owned the Skegness pier. It was a summer's day when I took what was the forerunner of the IBM PC in the back of my company car,

LEAPFROGGING WITH GENGHIS KHAN ON LEVEL FIVE

IBM 5120 Desktop Computer

a Renault Estate. The 5120 looked like a Russian Television set and was blessed with 32kilobytes of memory and twin eight inch floppy disk drives that held a princely 1.2 Megabytes of data each. The 5120 understood programmes written in Basic and a language laughingly called APL (Advanced Programming Language). The secret of my success however was what was described as a "Fourth Generation" Programming language called "BRADS" (Business Report Application Development System), a forerunner of the database programmes developed for the IBM PC that became a part of my business. BRADS allowed me to create a simple data file containing details of all the receipts from the legion of slot machines this company leased out in pubs and clubs all over Lincolnshire. The company management supplied me with some sample records and, ignoring the John Newbery demonstration philosophy, I showed them live how easy it was to create these records, sort the results and print an analysis report.

The whole exercise took less than an hour, including setting up on their boardroom table and was concluded by a celebratory visit to Skegness Pier and a free ride on their new "Virtual Reality" experience "The Jules Verne Astroglide". This amusement park ride was very advanced for its day but tame by today's standards. It resembled a NASA space rocket lying on its side. Once inside this rocket you were seated in a mini cinema that held about a dozen people with a screen in front of you. Once the doors opened you were treated to a simulated flight into space whilst outside Skegness Pier rocked as the hydraulics jerked and shook you to simulate gravity. It was all great fun and a happy conclusion to a memorable day.

Working for IBM was like being in a close-knit religious family in which members are urged to "Keep the Faith" even when evidence seemed stacked against your products. I met some fantastic characters within the company but there were also many who displayed a certain arrogance about "Big Blue" products and services and regularly used what became a familiar phrase "No-one ever got fired for buying IBM". Whether this was true or not was beside the point but this attitude did not fit comfortably with my own values. My one year training was completed with a two week nerve-jangling or as the Manchester United Soccer Team Manager Alex Ferguson would describe it "squeaky-bum" time that was the IBM Sales School. Rarely can anywhere else exemplify the rewards of success and the pain of failure than IBM Sales School. Success in Sales School meant that you graduated as a fully fledged IBM Sales and Marketing Executive and joined your other successful course mates in a celebratory meal at which you were presented with a tankard. Failure meant that you were sent home before lunch, either out of a job or relegated back to the ranks of Systems Engineer at your home branch.

These two weeks were an intensive period of assessed dummy sales calls and presentations to the tutors with continuous feedback and assessment so you did get warning if things were not going well. This practice brought me one of the funniest moments in my training when I was present in a call being made by a Bristol Salesman called Phil White, who bore an uncanny resemblance in build and appearance to the UK comedian John Cleese in his role as Basil Fawlty, the hapless Torquay hotel owner. Phil was an ex-OPD salesman who, on this final week, had been warned that his performances were borderline and that he was in real danger of failing unless he improved (no pressure there then). The previous day, Phil was given his briefing for the call which was to be made to our course tutor role playing the part of the Financial Director of an Export company. Amongst the details in the briefing were some information about size of company and the fact that the company's existing accounting was done on an ageing Olivetti ledger card machine. Digital native readers are likely to be totally bemused by this concept which amounted to a computer with main memory but no disk storage. Instead of disks, the account records were stored on magnetic stripes (like those on credit cards) fixed to a thin card. These magnetic strips held the account data for an individual client and were updated by the computer in a form of sequential processing.

In his desperation to make a good impression in the dummy call the following morning, Phil did all he could to do the necessary research in preparation for the ordeal that was to follow. He got some competitive analysis of the Olivetti machine to understand its weaknesses and he also brushed up on export companies who had purchased IBM, acquired from IBM's archives of successful case studies. Nobody could have worked harder than Phil the night before his call. I was in the room with another salesman to observe Phil work his magic and when it came round to his turn, he stood up confidently and tapped on the desk where the role-playing tutor was sitting. "Good morning – I'm Phil White of IBM and I'm here to see your Financial Director". The reply to this statement was as unexpected as it was devastating to Phil's carefully crafted sales plan. The tutor replied "I'm sorry Mr White but our Financial Director has been called away on urgent business – he asked me to see you instead. I am one of the office clerks here." Phil turned his head towards us as it sank towards his knees and he uttered the immortal reply in a pained voice "Oh Fucking Hell ...." We immediately burst into helpless laughter and the tutor could not help deserting his role playing character to join in our mirth before telling Phil that although he should have been obliged to give Phil a disastrous UDH mark, he admitted that the case study was somewhat unfair in these circumstances. He did give Phil another chance after reminding him that this sales call was based on an actual case study in which the real-life salesman had used the opportunity to win over the junior clerk and use him as a champion to the IBM case when the Financial Director returned. Happily, Phil did survive and went on to be a salesman.

After sales school I returned to the branch and my first six months with quota responsibility. By this time my Marketing Manager had been replaced with the legendary Dick Jeffries – a very capable cockney salesman who had risen through the ranks. Dick had some very engaging and distinctive mannerisms which included pointing both hands towards you as if he was holding two guns and saying "Spot on" in a strong London accent. It is one of Dick's many experiences in his salesman days that I have often used to tell people about him. It was an occasion at the end of his branch's quota year (IBM's ran from January to December) and his branch was getting very close to making quota but needed a sale to clinch it. Dick had for some time been trying to sell a machine (most likely an IBM System/32) to a construction equipment hire company called Thwaites. Each of his previous calls had followed the same circuitous fashion in which the MD recognised that he needed a new computer but would not sign because nobody in his company knew anything about how to use them. In a last ditch attempt to finally close the deal, Dick was accompanied by a senior IBM regional manager in support. As the call developed into its familiar pattern of argument, an increasingly frustrated Dick allegedly said "You needn't worry – I've sold these machines to thicker "tw*ts" than you". As the regional manager looked on in disbelief, the potential client burst into laughter and signed the contract shortly afterwards making Dick a hero and a legend in his own lifetime.

My first six months on quota did not prove as easy as the very first success with Wold Farm Foods and it was clear that, as December approached, I would be unable to hit my target as my pipeline of prospects was not as big or as advanced in the sales cycle as necessary. As it happened, the Nottingham branch and one of its other salesmen needed just one sale to make quota. I had been in discussions with a new prospect who was based in Leicester and was called Global Watches. This company was the UK distributor for Sekonda watches and had been formed and was run by a General manager called Keith Brooks. He had originally worked for the competitive UK Timex company and had been approached by the Russian manufacturers of Sekonda, keen to break into the UK market, to set a distribution company up for them. Keith had become quite friendly with me and had told me his amazing life story which included being put on board a steamship for South Africa at the age of fifteen and told to fend for himself. Sekonda made Keith an offer he couldn't refuse and he told me that it was like a licence to print money.

Although it was clear that Keith had recognised the need to buy a system to track his post-Xmas returns of faulty watches for repair or replacement, he was a very busy man and, in my judgement, not yet ready to commit. Dick Jeffries had analysed all the sales opportunities amongst his team and reached the conclusion that Global Watches, even as the

Xmas holiday started, was his eleventh hour hope of quota (and his bonus). As I have grown older it has become increasingly difficult to believe that I actually went to Keith's local pub on Boxing Day to get him to sign the contract. Being an ex-salesman himself, he was sympathetic to my cause and readily agreed, which made me a hero as I saved the branch and also helped a colleague to make quota. My reward for this act of self-sacrifice in the cause of my fellow men was £800 to spend on a two week holiday in Greece for myself and my first wife, Angela.

There were many eventful and memorable moments in my time at IBM. It was with this company that I learnt so much not just about how to sell and about the technology behind computing, but also about many different industries and about the sea changes that were beginning to occur in the world of the data processing department. The last minute rush for everyone to make quota by December 31st became an annual ritual, fortunately aided by many companies also wanting to spend their IT budget before the financial year end. I was involved in yet another amusing end of year incident when IBM had organised one of their "Executive Briefings" in Italy where many IBM systems were manufactured. These were prestige trips on board the IBM private Lear jet based at Heathrow and we were invited to nominate potential customers who were most likely to sign a contract after such a visit. By this time, my patch included Northampton and one of my prospects was a family book distribution company called "James Hadley". Dick Jeffries, accompanied by a Senior Sales Engineer (SSE) had been self-nominated to escort two company representatives on the trip.

I had been trying to sell John Hadley an IBM System/34 and recognised that they were close to signing so I nominated their him to travel with Dick and the SSE down to Heathrow for this prestigious trip. I was settling down to watch television in my Great Glen home when I received an urgent phone call the evening of trip. Dick told me that they had been relaxing at Heathrow waiting to go into the departures area when the SSE casually remarked "You know, the last time I went on one of these trips, one of my customers had forgotten his passport". John Hadley turned pale and said "Do I need a passport? I thought I wouldn't need one as it is a private flight in Europe". The upshot was that John

IBM System/36 Computer

Hadley had to be unceremoniously left to make his own way home to Northampton whilst the others continued on their journey. Dick asked me to call John the following day to make sure he wasn't upset with IBM. Fortunately for us he did see the funny side and agreed to sign the contract by the due deadline.

Northampton was also the location of one of my other sales conquests in the shape of the "Acme Clothing Company". It was a practice in the office for the secretaries to take messages in our absence and use little "Post-it" notes to stick on our desk with the name and phone number of the sales lead. My first reaction when I saw the name "Acme Clothing" was to suspect that one of my colleagues was playing a joke on me but when I was assured it was genuine I made the call and light heartedly commented that I wondered whether many of his customers thought his company might be associated with Bugs Bunny. It was clear from the silence at the end of the phone that this did not go down too well. In the event, he was willing to accept a visit from me after our brief conversation and I went down to Northampton to meet him on site. The Acme Clothing Company were distributors of alternative street clothing and it was their catalogue the first opened my eyes to the popularity of the "bum-flap" as a fashion item. This semi-circular piece of cloth was worn with a belt round the waist to dangle over the rear portions of their fashion conscious wearers. This company also turned out to be another unlikely "Bluebird" as they signed contracts for a new computer with little persuasion.

During this period in the early 1980s, IBM were making regular enhancements and additions to their product line. The IBM System/34 which had been so successful had an upgraded version called the IBM System/36 which had a massive 256 kilobyes of main memory against the S/34's limitation of 64 kilobytes. It was also smaller and I believe had more disk capacity and better terminal handling. We were assisted in our sales efforts by a number of local software houses who enjoyed something of an incestuous relationship with the salesmen. One of the most successful of these was a company called CP Programming Services whose Managing Director was called Phil Hatton. Most salesmen formed alliances with the most competent of these local companies and many later left IBM either to join a software house or set one up. The S/34 and S/36 both had their own proprietary operating system which looked after the essential systems aspects and left the software houses and customers free to programme new business applications using a programming language called "RPG" (Report Programme Generator) which, in many ways, was a bit like the BRADS software I had used so successfully in Skegness.

RPG was easy enough even for me to understand and so I began to use our own in-house S/34 in Nottingham to write my own programs to help me manage my prospects and customers. What I developed was really a simple database that not only kept the basic details about the company but also tracked my contact with them, and so managed the sales cycle. Meeting my sales quota became a regular annual success and was rewarded not only by annual salary bonuses for "OTE" but also by inclusion on the annual IBM sales conferences where salesmen let their hair down in an orgy of self-congratulation fuelled by much entertainment and alcohol. One of the conferences I attended was in Barcelona where the cabaret dinner included two performances by Andrew Sachs, one as his Fawlty Towers Alter Ego the hapless Spanish waiter Manuel and one in the guise of a German professor. Both of these were intensely entertaining and were supplemented by a hypnotist who succeeded in getting some of my IBM colleagues to do things on stage they might later regret.

Once the dinner was over and we all spilled out into the street to seek further excitement, Doug Hall and myself found ourselves in a little cafe where, amazingly, I was sat back to back with Andrew Sachs. He proved a very charming man who seemed only too willing to share his anecdotes about Fawtly Towers and the demanding attention to detail shown by John Cleese. After this encounter Doug was still thirsty for more adult entertainment and so persuaded me to follow him into a strip show where we were frequently "lobbied" by the artistes to buy them champagne – a euphemism, as I later found out, for more personal services. Doug seemed to be getting on well with one such artiste when she made her apologies and disappeared only to re-appear on stage as the next act. Doug became ever more enthusiastic as the layers of clothing were tastefully removed until there were only two fans to cover her modesty. As she teasingly moved the fans aside, Doug's face dropped when "she" revealed that her top half was all woman whilst the bottom was definitely all man !!

The years I was working for IBM between 1979 and 1984 signalled the dawn of mainstream business computing and were arguably the beginning of the consumerisation of the desktop computer. I remember on more than one occasion being mildly irritated by potential customers countering my suggestions for a solution by saying something along the lines of "Well, my mate down the pub says I ought be buying XYZ computer". It was tempting to enlighten the originators of such comments about the amount of training and experience I had of technology compared to "the mate down the pub". The irony about the changes that have taken place in the computer industry is that such remarks were ahead of their time and quite visionary in nature because today they would have a good chance of being true!! The days of computer giants such as IBM being able to afford the salaries of highly qualified salesmen to sell what amounted more often than not to basic financial accounting systems were drawing to a close to be replaced by "off the shelf" desktop computers that would eventually find their way into High Street retail stores and even supermarkets whose shelves in those days would only be stocked with food.

There is no doubt that solutions such as the S/32, S/34 and S/36 were responsible for much of the demise of offices full of typists and clerks carrying out the management of sales invoicing, credit control, stock control, purchasing, management accounting and nominal ledger. In this role they helped the small and medium businesses to become more competitive and efficient by cutting the human cost base. Perhaps it was this trend that saw the dawn of the age when machines began to use their memory, intelligence and senses to get the best out of human beings? In GSD, IBM had a relatively limited (by today's standards) variety of technology solutions. The focus was almost entirely on the accounting and business administration applications that represented the bulk of the demand. The other mighty mainframe companies of the day such as ICL, Burroughs and NCR became very vulnerable to IBM GSD products and solutions but GSD itself was beginning to see stiff competition in more niche markets from DEC (Digital Equipment Corporation) and in the GSD market from Hewlett Packard and Wang. I remember doing a sales call on a social friend from the squash club who ran a very successful label printing company called Labelsco. Bob Robinson was a friend and the Founder and Managing Director of the company. He had been a salesman previously with Rank Xerox and so knew very well the reputation of IBM. I could tell that he wanted to buy from me but his main application was for a machine that would help him design and print labels more quickly and cost effectively and the honest answer was that GSD had nothing designed to do this.

He was very surprised and disappointed because he felt sure that IBM must have a solution. At the time all IBM had was the Series 1, in many ways far less sophisticated than the S/32 and S/34 family but very flexible in its configuration and applications which included factory automation. There were very few software houses specialising in the S/1 in those days although an erstwhile colleague in IBM Nottingham, Ian Dawkins, later left the company to run a S/1 specialist software company called Krypton Computers. This was the destiny of many of my sales colleagues as they saw first-hand the riches to be gained by running an

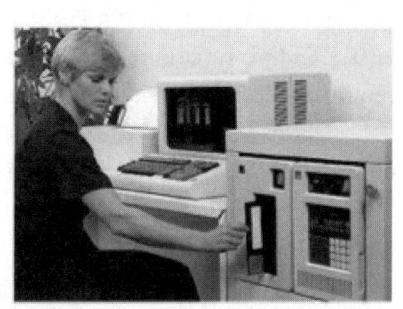

The IBM Series 1 computer

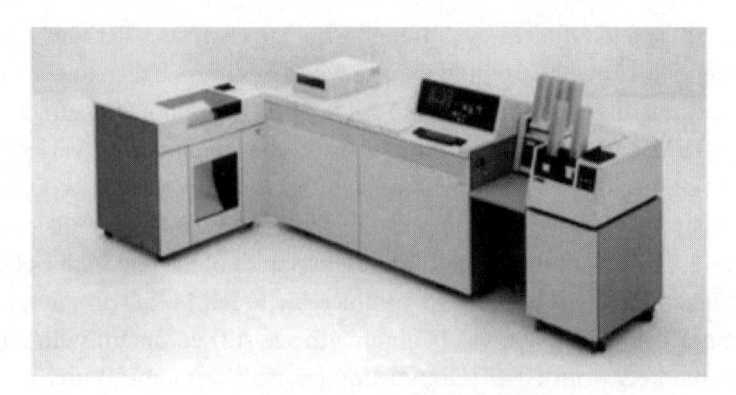

IBM System/38 Computer

IBM recognised software house. I did wonder if the mass recruitment of sales executives around 1979 followed by the mass exodus of the same people from 1983 onwards was in fact IBM's cunning masterplan to train and indoctrinate legions of individuals who would later leave the company payroll but continue to generate IBM sales without the salary overhead.

The product portfolio of GSD was about to be massively and significantly extended by the arrival of the System/38 whose name belies its origins. Far from being a natural extension of the System/32 family by offering more memory and capacity, the System/38 was the nearest thing to a visionary revolution in computing, and light years ahead of the competition in its concept and architecture. The S/38 was the legacy of a blue sky project in which scientists were given a free hand to imagine the computer of the future without any of the constraints of backward compatibility. The result was a 64 bit architecture that we are only just seeing appear in today's powerful desktop computers. The S/38 effectively had an inbuilt relational database that provided powerful and cost effective functionality yet, like its S/34 and S/36 cousins, did not require a special environment to function.

The S/38 proved to be a real winner for me because I had total confidence in its sales potential, especially competing against the dinosaurs that ICL, NCR and Burroughs had installed. It was largely through the S/38 that I became a NIU (non-IBM user) specialist salesman whose job it was to target and replace these ageing competitors in corporate accounts. One of the main barriers and arguments used against anyone trying to replace a competitive mainframe was that companies had so much invested in legacy programming and systems, it would be prohibitively expensive and time consuming, not to mention risky, to replace mission critical applications with a totally new solution. The productivity and low operational and skill costs that the S/38 offered won the day on many occasions with the exception of one notable situation that brought an unexpected competitor, IBM DPD.

I had for some time been pitching the S/38 to a UK car rental company called Swan National whose admin HQ was in Leicester. They had an old ICL mainframe which was very vulnerable to the IBM S/38 strategy. I felt that we were in with a good chance. The Data Processing Manager had told me that they were also looking at the HP 3000 mini computer which was more like our S/34 solution. It therefore came as a complete surprise when Dick Jeffrey and I learnt that our DPD counterparts just down the road were also pitching to Swan National with a solution based on the mainframe IBM/370, a much bigger computer but requiring traditional resources to operate it. Dick and I were duly summoned to City Gate where our equivalents there, a marketing manager and a female salesperson nicknamed "Thunder Thighs", made it quite clear that GSD "toys" like the S/38 were not up to the job and that they already had high level influence within Swan National through the then MD of IBM UK, Eddie Nixon. Both Dick and I had to withdraw graciously after putting up our arguments and it brought a certain smug satisfaction when Swan National eventually bought the HP 3000 and DPD were left with egg on their faces.

The S/38 was also responsible for securing me my first trip to Canada when IBM ran a national competition for the best System/38 presentation where the first prize was a one week "study tour" to take in Montreal, Toronto and Vancouver. The idea was to motivate sales staff to develop a better understanding of the machine's competitive advantage. The competition had one award for the north of England sales region and another award for the south. I was nominated to represent the Nottingham branch and went down to the soulless Basingstoke offices to strut my stuff. I finished up winning the northern section and Clive Jecks the southern. We flew together to Montreal on the first leg of the journey and spent two days at each of the cities on our itinerary, time which was a mixture of meetings with System/38 specialists and tourism. In Montreal we visited their famous Chinatown and the World Expo centre, and in Toronto we had a fantastic meal in the revolving restaurant of their skytower and a trip to Niagara Falls which was breathtaking and magnificent. Nothing can prepare you for the sight of this massive volume of water tumbling over the edge of the falls. Your mind tells you that the lakes and rivers that feed the falls must be running dry to sustain the incredible and powerful current.

My favourite leg of the journey was in Vancouver. We arrived at the airport late in the afternoon and our taxi took us over the bridge into the city just as the sun was setting over the distant snow capped Whistler mountain and I immediately fell in love with the city. We could see the lights twinkling in the distinctive green roofed buildings, the sea and the mountains all in one spectacular vista. We were lucky enough to have most of our last day free before taking the evening flight back to Heathrow so we hired a car to take us north through the forests to Whistler and on to an Indian reservation where I saw my first and only real live rodeo.

I made a number of S/38 sales as an NIU salesman, including one to the Hogg Robinson Group who ran insurance and travel operations from their offices on St Margarets Way in Leicester. Their IT manager at the time was John Kermode and he readily grasped the significance of the S/38 architecture. It was around this time that I was looking for opportunities to start my own business. I had just sold another S/34 to a company called Travelsphere whose offices were in a former school in Coventry Road, Market Harborough, a building which would later figure in my plans to set up the Harborough Community Learning Network. Their Marketing Director was Richard Mackay and he and his fellow directors were very ambitious and entrepreneurial. One of my dreams had been to set up a prestigious management training college with leisure facilities in a country mansion setting and around that time the former British Steel Training centre in the village of Gretton, not far from Rockingham Castle, came onto the market. John Kermode, Richard Mackay and myself spent some time exploring a joint venture to acquire the building set, as it was, in a beautiful rural spot overlooking the Welland

Valley. Our discussions eventually fizzled out but my connection with Richard Mackay continued at a later date through he and his wife Celia's involvement in Leicester's Little Theatre. Travelsphere became hugely successful and Richard was able to finance the development of a beautiful up market hotel in massive grounds at North Kilworth in Leicestershire.

In what was to be my final quota year in IBM, I got to look after the Courtaulds account in Nottingham and Derbyshire and built up a very close relationship with them to help plan solutions which would improve the efficiency of their supply chain management and make it possible for their factories to respond quickly to changing patterns in retail fashion sales. This resulted in a major order for IBM computers in the December of 1983 and helped me to achieve 200% of my sales quota for that year.

My social life with my Rentacrowd friends at Squash Leicester and my involvement with the Kibworth Drama Group occupied my spare time and it was through one of my Rentacrowd friends, John Porter, that I got the opportunity to start my own business in the way that I had dreamed of in the boredom and tranquillity of Cooden Beach. John Porter was the founder and owner of Venture Business Forms where Sue, the wife of my best mate Dave Timson, also worked. Their offices were above an opticians in London Road Oadby and they were doing their accounts on an old ledger card accounting system. John was ambitious, clever with his investments and also status conscious. We discussed the benefits of the S/34 to his business and struck the deal that led to the setting up of Mass Mitec. I managed to get hold of a second hand S/34 at a good price and, in exchange for the savings John was making compared to brand new machine, he agreed to give me free office accommodation and secretarial support plus a £6,000 lump sum of start up finance. This, together with the commission I had earned from the Courtaulds deal, was enough to get me started and so I gave IBM three months notice that I was leaving.

It was a great credit to IBM that they continued to support me during those first three months of 1984 that I served out my notice, including allowing me to attend the annual sales conference in Nice, France only 2 weeks before I finally left to start the next chapter. At the time I was living in Oadby's Forest Rise, not too far from my new workplace. Venture Business Form's office location provided me with the first

An early IBM PC with monochrome display and 10 megabyte hard disk drive

challenge of my new business by virtue of the narrow staircase that led to their offices. Its restrictive size made it impossible to get the S/34 into the office via the stairs and so it was that we had to hire a crane, take out Venture's main office window and organise police traffic control one Saturday morning. In a smoothly co-ordinated plan, the S/34 was hoisted into the air and swung through the empty window frame into the office before everything was restored to normal and I set the S/34 into operation and trained the staff in its use.

The £6,000 that John had provided gave me the capital to invest in my own computing technology and prove that the new era of computing for competitive advantage through innovation had arrived, even for the smallest micro enterprise. I bought an IBM PC with 64 kilobytes of main memory and two 5 inch floppy disk drives each of which held 320 kilobytes of information. This standard system with its monochrome monitor displaying light green text and a very dark green background cost around £3,000, in financial terms probably equivalent to £30,000 at the time of writing this in 2011. It was a massive relative investment for a start-up with no track record and no experience of running a business.

I called my sole trader business Mass Mitec which stands for Marketing Administration Support Systems and Marketing Information Technology Consultancy, quite a mouthful but the name served me well

Rolodex Card Holder

as it gave the impression of being a larger organisation. I became D J Wortley T/A Mass Mitec and set about trying to get some customers. I hired an accountant who advised me to start trading on April 6th to give myself the maximum period before I needed to pay tax. I learnt very quickly that it takes much longer than you think to bring business in and that when it does start coming in it is usually not from where you most expected it. The desktop computer in 1984 was very basic but I was determined to follow the practices and disciplines of managing my prospects and sales leads that I had developed in IBM.

I had set up a kind of Rolodex card system containing the names, addresses, telephone numbers and contact details for as many of my old IBM contacts as I could and I set about sending out letters telling them of my services. I later designed my own company logo with help from one of my friends to provide a high technology feel that reflected the nature of my business which was originally intended to be a strategic consultancy advising on how to use information technology to generate new business as well as save costs.

The software available for the IBM PC in those days was very limited and quite expensive. I bought myself a word processing package called Multimate and Lotus 1-2-3 spreadsheet software, both of which were becoming industry standards. I was constantly looking for innovations that could help me to promote my services more effectively but this was in the days before the Apple Computer and Microsoft Windows. The IBM PC had an operating system called DOS (Disk Operating System) which involved typing text commands into the screen for everything including loading programmes and copying files. The IBM PC in those days had no graphics and no hard disk drive. Its basic main memory for running programs was 64 kilobytes and Bill Gates had once famously said that he could not envisage the day when anyone would need more than 640 kilobytes of memory to run programs.

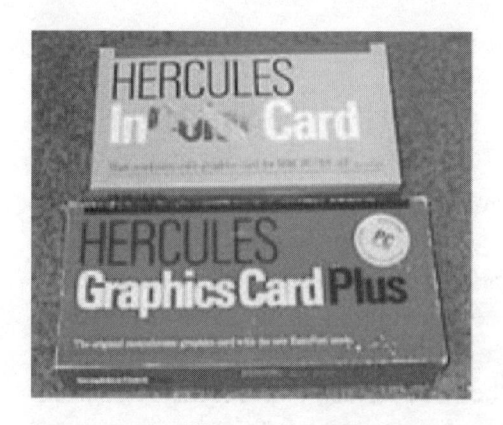

Hercules graphics cards

I was constantly adding more capability to my PC and this included a dot matrix printer from Mannessman Tally that could not only print out my reports and accounts but could also do letters and mailshots. This was a period in my life when I really "got my hands dirty" upgrading and maintaining both Venture Business Forms S/34 and my own desktop PC. I added more main memory to increase the capacity to 256 kilobytes which involved taking the printed circuit board out of the machine and plugging in additional memory chips. I added a 10 Megabyte hard disk drive which came on another circuit board that slotted inside the PCs case and as soon as it became available I bought a "Hercules" graphics card that would allow me to "do graphics" on my monochrome system.

I set my PC up with the Braid Telex Manager and that allowed me to do telex mailshots to my marketing database as I started to organise seminars to attract customers. I was looking for a database software package that could give me the kind of relational database tools that I had seen on the IBM S/38 and that I wanted to adopt to create my own marketing application. There were database packages like Delta available but almost without exception they were simple flat file systems that were good at keeping basic records but very limited if you wanted a system that could relate details of transactions to customer records. Eventually I discovered a package called RBase and I set about using it to manage my marketing database. I used Rbase for many years afterwards and was able to program my own applications such as the management reporting system I was to use in later years for our Prontaprint national presentation network.

The early months after I started Mass Mitec were a real mixture of consultancy, programming work and the occasional sale of pieces of hardware or software. Courtaulds were a very important client in those early days when I was pioneering the use of desktop telecommunications for both telex and electronic mail. It was a chance visit to a computer exhibition at Birmingham's National Exhibition Centre that set a more strategic opportunity for the business. As I wandered around the stands I saw 2 different exhibitors who together would provide me with the foundations of Mass Mitec Presentation Services. The first of these companies was offering a PC program called Execuvision that allowed you to create your own visual presentation material to display on the recently launched IBM PC colour monitor. The resolution of these early monitors was only 320 x 200 pixels and the maximum number of colours on any one display was four. The quality by today's standards was appalling.

On a completely separate stand Polaroid were showing off their brand new Polaroid Palette film recorder that you could connect to your PC to make 35mm slides for business presentations. It worked by effectively taking a picture of whatever was displayed on your computer screen and was plugged into the monitor port at the back of your PC in between the computer and the monitor. This chance stroll round the NEC to see what was new helped me put the pieces of a new jigsaw puzzle together that would eventually establish Mass Mitec as one of the UK's leading and most respected presentation graphics companies. The full story and history of my involvement in imaging technologies is covered in depth in the next chapter but this connection of graphics software and a 35mm slide making device was what I saw as allowing me to make more professional presentations without having to pay expensive design agencies or buy a very costly graphics workstation such as the Quantel Paintbox which in those days would have set me back over £100,000 for capabilities which today anybody can find on the cheapest laptop computer.

In the first year of operation my business had begun to develop to the point where I was able to recruit my first member of staff, Sue Berry, to help me with office administration and sales activities. The S/34 at Venture Business Forms was running smoothly and it was time for me

Compaq portable computer

to move into offices of our own. Almost exactly across the road from Venture Business Forms was the Chris Sharpe Motorcycle shop with a hairdressers above it in this two storey building and it was here that a new episode in my life was soon to begin.

As my business evolved in our new home, IBM began introducing new models of the PC with more capacity and functionality. The IBM PC-XT was announced with its 10 megabyte hard drive to replace one of the floppy disk drives and the IBM PC-AT followed this and used a faster 8086 processor instead of the old 8088 processors. They also introduced the colour monitor and graphics card that could display a maximum of four colours with a choice of different colour palettes. As time went by we acquired an IBM Compaq portable from one of our clients. The word portable was misleading because they were very heavy to carry and shared the size, shape and weight of the sewing machines of the time. It was in 1985 that the first CD-ROM disk drives began to appear and another five years later in 1990 that recordable CD technology for the PC was introduced by Phillips.

In the years to follow, we also bought some of the early Amstrad computers as they became available and were far cheaper than their IBM equivalents. As our presentation graphics business grew I was selling specialist hardware and software to our clients. I had soon learnt that the majority of customers were far more interested in setting up in-house solutions like the ones we were using at Mass Mitec. It was comforting to know the blue chip clients like BT and National Girobank looked to a small company like ours to improve their own facilities. We set up accounts with the main software distributors like P&P Micros in Rossendale and Softsel in London. By this time my early worries about where the next piece of business was going to come from had been

Amstrad 1512 PC

replaced by a growing realisation and confidence that as long as we continued to do all the right things in serving our customers, the revenue would surely follow. Today, the business environment is totally different. Whereas for decades, perhaps even centuries, trade relationships which had been built up over time into strong brands could be relied upon to create a relatively stable world, new competition can now spring up overnight and new developments can challenge even the most established of companies.

I have always believed in the value of lifelong learning and investment in keeping up to date with the latest technology and one of our main suppliers, Softsel, ran a tremendously valuable annual event at the Renaissance Hotel next to Heathrow Airport. The two day event was called Softeach and gave hardware and software resellers like ourselves an opportunity to listen to hourly seminars of our choosing from all the major solution providers. I attended every one of these events and would almost always bring my staff along as part of their education and team bonding. Softeach vendors also offered lots of free software and hardware in prize draws which we were lucky enough to benefit from on several occasions.

I continued to develop our own in-house administration systems to constantly improve our workflow processes. I designed and implemented the Rbase database that we used to record and report on our growing presentation services business with Prontaprint and, after the settlement of our insurance claim for the burglary, I was able to afford to have a local area network installed and we bought a powerful HP server with Microsoft Exchange Server. This gave us the chance to create a central shared contacts and calendar system that everyone in the office could access. As the years rolled by towards the late 1990s, the power balance was beginning to shift. Setting up, upgrading and maintaining our hardware and software was getting easier and easier and the experience, intelligence and senses that I had needed and that my clients had been willing to pay me for were called into play less frequently. Whereas once a corporate client would need our expertise to both design a presentation and produce professional quality output, they were now able to use Microsoft Powerpoint and a data projector or one of the high quality low cost desktop colour printers that were entering the market.

Like many others in business, I was becoming more reliant on the computer's memory and programmed intelligence to help me fulfil my potential instead of the other way round. These changes and improvements in the capacity, power and processing capabilities of computers helped to speed the growing pace of "disintermediation" or removal of the middle man in commerce. End users no longer needed the wisdom of the experienced salesman or the purchasing power of distributors. This end user access to powerful software and hardware that was once the domain of the IT professional led to the situation where even my milkman, without any formal training, could achieve better results than my years of industry knowledge was capable of. Even my Uncle Len, whom I had given a second hand computer with Harvard Graphics installed soon after he had retired from his job at the Ford car plant in Eastleigh, was showing me shortcuts in the software that I had blissfully been unaware of. As more and more powerful computers appeared in retail stores like PC World, the home computers rapidly began to overtake their office counterparts in power, capacity and software functionality and it is still often the normal state of affairs that people have more powerful computers at home than they use at work.

As the chapter of my extensive involvement in using my skills to get the best out of computer technologies drew to a close with the termination of my involvement in the Harborough Community Learning Network that had been my invention and my passion, I found myself in severe financial difficulties and unable to make the shift back into employment. I became one of the victims of the middle age scrap heap and, to avoid losing everything I had worked for, desperate measures were called for. This is where the story that spawned the title of this chapter originated and brought home to me in a very salutary way how much power had shifted from man's intelligence to the machines he invented.

I saw an advertisement in the Leicester Mercury for "Warehouse Operatives" at the Tesco distribution centre near Daventry, about twenty minutes drive from the White House in Lubenham. The job involved working 12 hour night shifts for four consecutive days then four nights off in a continuous cycle. The salary was around £21,000 a year but would at least give us a regular income that could pay the mortgage and still allow me time during the day to continue my consultancy work. I did not consider that working in warehouse was in any way not worthy of my talents in these circumstances and so I applied for the job. When I went for my interview I was surprised at how rigorous it was. As well as having a medical examination, all the applicants were given a comprehension test to assess our numeracy and literacy skills and we were even divided into teams tasked with building a "Lego" bridge sufficiently big and wide enough for a team member to crawl through it without knocking it over. Each of us was also asked to tell everybody something unusual or especially interesting about ourselves. This was at a time when I was able to do the odd bit of TV extra work, so I impressed everyone by telling them I had been on TV in programs like Doctors. In future this would lead to the odd occasion when someone at the warehouse would stop me and say "I saw you on telly last night". Fortunately I was successful and, on the first evening shift, collected the padded thermal ski suit, gloves, balaclava and boots that were to become my uniform.

The work area of the warehouse operated at a temperature of around minus twenty degrees Celsius. The staff who carried out the work were a mixture of full time permanent employees like myself supplemented by sub-contracted staff who were bussed in each night in numbers that reflected the workload that night. The majority of these contractors were foreign and seemed to be a mixture of refugees and asylum seekers. Every working night I would drive to Daventry and park my car in the car park before going through security into the main warehouse where we would clock on and then go to the locker room to change into our ski suits for the evening's work. Everyone would then gather in the huge staff canteen to hear the job allocations for the night given out by the team leaders. This allocation of jobs was preceded by a review of the previous night's performance against company targets. The warehouse essentially received bulk frozen food from manufacturers like McCains, Walls and Birdseye. These arrived in lorries whose palettes were unloaded, checked against the original order before being automatically routed to a part of the complex known as High Bay, where the palettes would be assigned to individual bays of this five storey building with five separate sections, A-E.

The orders received from the retail outlets would then be co-ordinated by the warehouse management system and used to get the right quantities of each item picked in High Bay and automatically routed by a complex conveyor system to the chute where that store's order was to be loaded into the familiar large trolleys we see being taken into our supermarkets from the Tesco lorries. Once each trolley is filled at the chute, it is collected by a special fork lift truck and taken to the appropriate bay where the lorry trailer for that particular route was backed up. This process meant that there were a number of different skills you had to be trained for and a selection of jobs you could be allocated on any one night. These jobs were High Bay, Goods In, Loading, Fork Lift and the universally dreaded "End of chute". The process of each team leader shouting out the names of those allocated to his section took me back to my school days when the captains of each playground soccer team would choose in turn who they wanted in their team. If your name was last to be shouted out it meant nobody wanted you.

My "favourite" job, and the source of the title of this book chapter, was High Bay. Every night I would hope that this would be my destination for the next 12 hours. Unfortunately it was also a first choice for most other warehouse workers, not because it was any less physically demanding but because it was a space you could be left to work alone with your thoughts except in periods of heavy demand. Being chosen for High Bay meant joining your team leader at the door to this building and being allocated a section. If you got A5 for example, it meant that you had to go to section A and climb up the five flights of stairs to the top level where you would swipe your identity card to let the computer know you were ready to start. Each section bore a grim resemblance to television portrayals of prisons such as the one Ronnie Barker found himself in during each episode of the series "Porridge". There was a central area which housed a conveyor belt and down each side where you might envisage cell doors there were openings in which the frozen food palettes stood waiting to be stripped of their contents. Above each palette bay was a red lamp and a small numeric display.

No sooner had you swiped your id card on arriving at your section than you would spot the first of the hundreds of red lights that would control your actions through the night. The single red light would show you where your first "pick" of the night was required and you would walk to the bay where you would see a number between 1 and 8. This was your instruction to pick that number of boxes off the palettes, turn and place them on the conveyor belt that was sliding along behind you. As soon as you had picked the required number your red lamp would extinguish to be replaced by the next location with its required number. Often, if the number to be picked for that order was greater than 8, you would have stay in the same bay until that order was complete before looking for your next red lamp. The sequencing handled by the computer ensured that you would spend the night walking in anti-clockwise circles around your area of High Bay until your labours were complete.

It was here at the age of fifty-four that I was confronted by the stark truth that after a lifetime of using my intelligence to get the best out of computers I was now under the command of a computer tasked with getting the best out of me and with a capability to organise the complex logistical activities of this operation way beyond that of any single human being. It was the computer that decided the optimum location for palettes, what needed to be picked and when and where it needed to be picked before deciding where it should be loaded and at what time. An army of intelligent human beings would not be able to do what this computer system could do and certainly not at the speed and accuracy of this sophisticated machine.

As well as within the ranks of the contractors, many of whom would not look out of place in a gangster movie, there were many foreigners amongst the permanent workforce, some of whom had started out working for a contractor before showing enough capability, reliability and enthusiasm to warrant a permanent job. One such Turkish guy was called "Ghengis Khan". He was a charming man who gave no indication of any propensity to rape and pillage. During the times when we happened to have adjacent lockers he would tell me his story about how he was a restaurant manager in Turkey before falling in love with and marrying an English girl. The warehouse job was all he could get despite his experience in other fields of work. This sad state of affairs was also true for a significant number of fellow workers who, through no choice of their own and despite substantial expertise, found that warehouse work was the only employment accessible to them. This situation made me think that we were the frozen foods industry equivalent of the "Dirty Dozen" – a bunch of hardened misfits and outcasts sent out on a mission impossible every night.

During my time at the Tesco warehouse I picked almost half a million boxes of frozen food in High Bay at an average of around 4,500 boxes in each of my 12 hours shifts. Under normal circumstances only one person would be required in each section of High Bay and as long as you carried out your sequence of picking and walking briskly the night would pass quickly, broken as it was by a 15 minute tea break every 2 hours, and a thirty minute meal break served in the canteen. On nights of heavy demand on specific sections of High Bay there was a need for a second person to cope with the workload and a change in the method of working to adapt to this additional worker. Normally you would only press your completed button after you had picked all the required boxes, but with two people you would both go to the first bay and press the completed button before picking in order to trigger the next red light in the sequence where your colleague would repeat the process in order for you to know where to go past him to. This process was known as "Leapfrogging" and is the reason for the title of this chapter. I worked at the warehouse with Ghengis for many months but our paths unfortunately never crossed in High Bay and so I could not fulfil the fantasy of "Leapfrogging with Ghengis Khan on Level Five".

My next preference after High Bay was the fork lift truck which had its forks at the rear of the truck instead of at the front. These trucks were just below chest height and had handlebars like my old Li150 Lambretta scooter but instead of a seat you had to stand upright on a platform. In order to be able to operate these machines safely, each worker had individual training which was concluded with a driving test in a separate part of the warehouse. This test involved steering round a course of cones backwards, collecting some trolleys on the forks and driving to another point. Once the instructor was confident of your ability you became eligible to be picked for live action. Wireless communication was used in the warehouse to send instructions to each fork lift. Your job was to constantly drive round the warehouse seeking filled trolleys, whose bar codes you would scan, and the computer would then transmit instructions to the driver on where to deliver these trolleys for loading on to the waiting lorries.

The only real downside of driving these trucks was the cold. In all other areas of operation the combination of ski suit and your heavy manual lifting was sufficient to keep you warm. On the fork lift you were not only standing still on this truck platform for long periods, you would also suffer from a "wind chill" effect from the speed of the truck. A balaclava or a good quality thermal hat was essential for this work but even these could not prevent a regular freezing of my moustache with icicles hanging over my mouth making me look like Omar Sharif's character in one of my other favourite films "Doctor Zhivago".

Next in the pecking order of tasks was "Loading" which involved working in pairs to drag the laden trolleys in the correct sequence into the back of the lorries ready for delivery to the stores. At the back of the warehouse were the loading bays where trailers would be backed up to the large warehouse doors by drivers called "shunters". There was a mechanism by which the connection between the trailer and the warehouse would be effectively sealed to prevent temperature problems. The store trolleys with their bar-coded labels indicating the store destination would be parked in individual bays according to the label so that all the trolleys for a particular store were in the same place and all the stores sequenced correctly in the order in which the deliveries would take place. As with all the other jobs in the warehouse, with the exception of the dreaded "End of Chute", the instructions for what to load, when and where were given on display panels by each lorry bay and each operative would indicate completion of the loading and activate the closing of the doors to once more seal the warehouse.

Loading was physically hard work and some of the lorries were so-called double-deckers which involved one person inside the lorry strapping the rows of trolleys in to prevent the load shifting during the road journey with the other person dragging the trolleys onto a hydraulic lift to be raised to the top level. I am certain that most people have no idea of the sophistication of the computer processes involved in getting the food and other goods into the shops. "Goods in" was not so demanding physically and, like High Bay, meant that you could pretty much work on your own. As the manufacturer's lorries offloaded their palettes into the receiving bays they were taken by fork life to be deposited on a track system that would present each palette in turn to the operative. The computer system would know from the bar code scan of the palette what goods were expected and the Goods In operator had to check the goods and attach another bar coded label to the palette so that the computer would know which section of High Bay was free and ready to accept the palette. The routing of the palettes to their final destination was entirely automatic and carried out by robotic cranes in High Bay. The main problem with Goods In was the cold temperature accentuated by the lack of movement through the night.

The worst of the jobs was certainly "End of Chute" and was normally manned as much as possible by the temporary contractors during each shift. The goods which had been picked and placed onto the conveyor belt in High Bay were routed automatically according to the product bar code before being sent tumbling down one of the dozens of chutes that led onto sets of metal rollers to the correct section containing the store trolleys for that order. There was an optimum stacking protocol based on the type of frozen food category so that you could keep vegetables separate from ice cream and other products. As the boxes would come sliding down the rollers in relentless procession the job would involve completing the trolley fulfilment before dragging it into a central aisle for collection by the fork lift truck drivers.

There were other warehouse tasks to cater for special circumstances like emergency orders and an area set aside for more manual filling of store trolleys. This was also a popular job because it combined fork lift truck driving with picking from palettes. In this role you would load a fork lift truck up with empty trolleys and be directed to different palettes of bulk goods where you would be instructed to pick the required number of boxes and load them on to the correct trolley. You would then spend the night stopping and starting your truck as you went round picking these smaller or non-standard orders until the trolleys were ready to be dropped off near the correct lorry bay. All of these activities were carried out to the accompaniment of whichever rock radio channel was favoured by the manager in charge for the night. My favourite track during my stay in the job was "I think I'd better leave right now !".

There were too many memorable people and incidents to highlight in this book but one particular memory does stand out in my mind. It was a night when the processes had gone smoothly and the workload had dropped in the hour before the close of shift. I had been working on "End of Chute" when my team leader gave me a huge sweeping brush and asked me to sweep up all the rubbish from around the chute area ready for the day shift coming in. I reflected on how far my career had come from my Post Office Telecomms scholarship, Distinction in postgraduate Management Studies, successful sales career and multiple business awards to now sweeping the floor of a warehouse. These reflections were made the more poignant because, once I got home after the shift, I had a quick shower and made my way down to Heathrow to catch the plane to Moscow where I was due to make a keynote presentation at the international eLearn Expo conference and would meet, amongst other dignitaries, the Chairman of the Russian Government eLearning strategy committee and Rector of Moscow State University for Economics, Statistics and Informatics, my friend Vladimir Tikhomorov. If only they knew what I was doing last night I thought to myself as I began my presentation the following morning.

The computing technologies employed in organising the little army of human beings in Daventry, working like ants in a colony or what to me felt like being a hamster tirelessly running round a wheel whilst I was in High Bay in order to put food on the shelves for my fellow man, gives an indication of how far these machines have developed within the span of my lifetime. Like every other machine that human beings have invented and then refined, we have successfully made computers that are smaller, bigger in capacity, faster, more powerful, more reliable and cheaper to operate. The consequences of our efforts are revealed in the architecture and components that function inside these machines but it is not the numerical magnitude of these improvements which has such unprecedentedly serious consequences for our race, it is the intelligence, senses and memory capacity that will be reflected in the following chapters that we need to understand quickly before it is too late.

The "fall from grace", back to basics experience I had in that cold Daventry warehouse in many ways brought me back into the human race and made me physically fitter, more philosophical and tolerant to my fellow man. Working alongside other men and women whose stories of hardship and injustice made my own loss of my business seem almost trivial was a positive part of my life for which I am grateful. In 2005 I saw the advertisement for the Project Manager post at De Montfort University and my journey through technology took another turn as I went from sweeping warehouse floors to being appointed as Director of the Serious Games Institute within 15 months and resumed my quest to use technology to make my own world and that of my fellow man a better place.

That early Ferranti Mark 1 computer, with its glowing valves and electric circuits and army of technicians in the year of my birth, had a very limited memory and calculating capacity. Twenty years later, computing had developed sufficiently to be able to put a man on the moon. The power, capacity and speed of the computers on board the Apollo spacecraft as it made its historic journey would, another twenty years on, be dwarfed by the tiny computer that plays music when you open these novelty greeting cards that cost just a couple of pounds. My own very first IBM PC had 64K main memory and 640K disk storage at a time when Bill Gates of Microsoft confidently asserted that he could not see a time when anyone would need more the 640K of memory. He was so sure of this fact that his Disk Operating System DOS was designed with this limitation in mind and yet he has become one of the richest and powerful men in the world.

At home I have a computer which has 32 Gigabytes of main memory and one Terrabyte of storage. I am typing out this book on a Dell laptop which has 4 Gigabytes of memory and 450 Gigabytes of disk storage. In my digital camera I have a tiny card the size of a postage stamp that can hold 16 Gigabytes of photos and videos, and my mobile phone has similar capabilities. My time at De Montfort University as Project Manager, and the wonderful four years I spent setting up the Serious Games Institute are covered in depth in the chapters which follow, but there was one experience in my time at the SGI which saw me draw parallels with the Daventry frozen food warehouse. It was when I was as guest visitor at the Dell factory in Northern Ireland and saw first-hand how this very successful manufacturers built the laptop I am using now.

As was the case with the Daventry warehouse frozen foods, very few of us ever think about what is involved in getting consumer items into our shopping baskets. We know and accept that our frozen peas were at some time growing in field somewhere in the world but we don't expect that they would be the same as if we had been picking them ourselves as I used to do in those summer Lincolnshire fields a lifetime ago. By the time I joined IBM, specialised and trained humans like myself and my systems engineers were using our knowledge in our dialogue with our customers to configure the computers to meet the exact needs of our clients with the correct hardware and software components. Especially on those VIP trips to the IBM manufacturing facilities we would proudly tell our potential clients that each machine is individually made for them. This was in reality only a half truth since the range of memory and disk options within IBM GSD was so small, most customers got what amounted to a stock machine. They could have any colour they liked as long as it was IBM grey !!

In the thirty years that have elapsed since those IBM days, manufacturing processing in industries like automotive and computers have become so sophisticated that the man in street lay person can actually configure his or her own machine to a very high level of detail that would have been hard to believe in the early 1980s. Within the Dell operation the machine that you choose the options for on their web site is genuinely individually assembled for you. All of the components required to build the laptop or desktop of your choice are individually assigned to your order and, in a remarkable operation, human beings actually put your machine together on a complex set of assembly benches before your final brand new laptop or desktop is boxed up with your address on the label and delivered to your door.

I have tried to reflect in each of my chapters mankind's achievement in making each of these technologies faster, more powerful, with greater capacity and cheaper. The chart below gives a rough guide to the extent of the success achieved in computing technologies from the period 1979 to 2011 which equates to my lifetime's main involvement in computers.

Year	System	Main Memory	Disk size	Price	Seller	Price /Performance Ratio
1979	IBM S/32	32KB	27MB	£27,000	Manufacturer	1
1984	IBM PC	64KB	640KB	£3,000	Dealer	0.43
2009	Dell Laptop	4GB	450GB	£1300	Manufacturer	43,269,231,000
2011	HP Desktop	32GB	1TB	£800	Retail Shop	1,250,000,000,000

These figures take no account of the inflation that has taken place over these years and that would make these figures even more staggering than they are. Where else might you see over a 1 trillion percent improvement in price performance within just over 30 years?

Amazing as they are, it is probably the architecture and infrastructure changes that have taken place over the same period which have even more significance to the future of mankind.

Year	System	Type	Processing	Data Storage	Control
1949	First Mainframe	Single User	Central	Central	Central
1970s	Mainframe	Single User	Central	Central	Central
1977	S/32 Minicomputer	Single User	Central	Central	Central
1979	S/32 Minicomputer	Multi-user	Central	Central	Central
1984	IBM PC	Single User	Local	Local	Local
2000	Lan Computing	Multi User	Local	Local/Central	Local/Central
2006	Thin Clients	Multi User	Local/Central	Central	Local/Central
2011	Cloud Computing	Multi User	Central/Local	Central/Local	Central

It is a few short years since my first real involvement in computers at a time when, like new born babies, they had little memory and processing capability and had to be communicated with in very simple language. Today, all mankind's intelligence and ingenuity has seemingly been embodied in almost unimaginably powerful systems that exist in a massive ethereal infrastructure we call cloud computing.

The burning question is "Have we created a God in our own image that we now worship and become increasingly reliant on to bring us our daily bread?"

CHAPTER FIVE

Oh what a Picture, what a Photograph!

This chapter on imaging technology relates to mankind's progress in using technology to capture, manipulate, reproduce and communicate pictorial information. It is this technology which has generated the majority of my income in the periods when I ran my own business. Our ability to photographically capture and print pictures of the world around us began in the 19th century and led to a demand for the skills needed to professionally take photographs, process the film and make prints and slides from them. Before she was married, my mum worked at a professional photographer's shop right next to the famous Boston Stump Parish Church and it was therefore quite natural for there to be quite a few photographic records of me as a baby, ready to be embarrassingly produced for guests later in my life. The Kodak "Box Brownie" was the consumer camera product in those days when all pictures were black and white although some photographers offered a special colouring service in which photos were "touched up" to make them look like colour photographs. Pictures were taken and stored on rolls of film inside your camera until you had used all the 12 or 36

Me impersonating a glove puppet with mum and dad in 1949

exposures and then you would rewind the film back into its cassette inside the camera, open the back of the camera, hoping that no light had got into the cassette, and then take the cassette into Boots chemist where it would be sent to a processing lab and you could collect your prints about a week later.

I vaguely remember having my own cheap Kodak camera at the end of the 1950s. To take pictures you had to open up the back of the camera and put your film cassette into one end of the camera and carefully pull out enough film to allow you to align the holes in the film strip with sprockets on a roller at the other end of the camera before shutting the back and using a little awkward pop up handle to wind the film on till you saw the number 1 in a tiny window on top of the camera. I lost count of the number of times that I either didn't fit the cassette into the camera properly, or the winding mechanism failed, preventing me from winding the film on to each new exposure, or I accidentally opened the back of the camera whilst the film was still unwound and ruined the film. All these mistakes led to numerous double exposures and pictures that looked as if a solar flare or nuclear explosion had struck just as I had taken the picture. This is a great shame as I would love to have had some of my early pictures of trains and our family life. I was not particularly good at taking photographs in those days and regularly managed to chop people's heads off through my imprecise use of the tiny viewfinder in my cheap camera. I did however manage to get a picture of Tina Turner on stage at the Boston Gliderdrome soon after she had recorded the classic Phil Spector production of "River Deep Mountain High" with her husband Ike Turner.

I can't recall exactly when colour films became available and affordable. Most of the pictures I took up to and including my time at Birmingham University were still in black and white but I do remember a colour photograph of me and one of the catering staff in the grounds of the university halls of residence. She was a Portugese lady whom I was just on passing terms with in the Halls of Residence and I think she wanted to send a photo back home to pretend she had an English boyfriend. After I left University and started work in Leicester, it was really when my wife Angela and I started our ski holidays that I began to take more photographs, mostly inspired by the beautiful mountain scenery that I so love to this day. The photographic print industry also now offered lower cost colour print solutions in which you sent off your film in the post along with payment and received back both your prints and a new role of film. This time in the early 1970s also saw photographic specialist companies such as Jessops of Leicester offering their own colour print services with low cost and faster turnround.

It wasn't until I had joined IBM that I had my first exposure to computer graphics. These were the very early days of computer aided design (CAD) with software for mainframe computers being offered by companies like Autocad to help create engineering drawings on monochrome screens. I had already had some experience with engineering drawings in my time with Post Office Telecommunications. In 1967/8 my student apprentice cohort all went on an engineering drawing course at Dollis Hill London where one of our number played out a "candid camera" type prank to observe human reaction to the sound and apparent sight of someone vomiting from a first floor window. This was only a simulation carried out as people streamed by below the drawing office window for their lunch break. One of our number filled a large glass with water and tipped it out of the window when no-one was immediately below its contents whilst someone else mimicked the sound of a "chunder" (vomit) as the water splattered on the pavement below. The effect was instantaneous and people started walking in a horseshoe path around what they imagined was something rather unpleasant.

One of my posts at Post Office Telecomm also involved responsibility for Drawing Office staff. This was an extremely skilled job carried out with pen and paper on large drawing boards. These complex drawings of equipment layouts and cabling were also transferred to what looked like tracing paper and at some point the drawings were photographed to be stored on Microfilm, a technology which allowed you to view these drawings through a magnifying device and also have them reprinted. All of this was very time consuming and probably unimaginable to anyone born after the invention of the internet. All the computer graphics at this time were in two dimensions (2D) because the processing power that we take for granted today did not exist even in powerful mainframes.

After I left IBM in 1984, my main interest in graphics was simply to be able to make my own presentation material. In those days you needed to go to a specialist advertising or graphics company to have your presentations drafted by hand and turned into slides using what was known as a rostrum camera – a special rig for mounting a 35mm camera in a clamp to get a perfect image of what the graphics artist had drawn by hand or typeset placed at the base of the rig. There were also at that time specialist graphics workstations such as the Quantel Paintbox but these were very expensive and only affordable by large specialised agencies.

My long journey in computer graphics began at that visit to the Birmingham National Exhibition Centre computer exhibition in 1985. IBM had just launched their colour monitor and graphics card offering four colours from a palette of 16 and a screen resolution of 320 x 200. There were a limited number of preset palettes you could choose from such as white, cyan, magenta and black. It was at this show that I saw a software program called Execuvision imported from the USA. They demonstrated how you could design your own computer graphics which not only included text but also cartoon graphics clipart. In my mind at the time it was an amazing development and when I saw the Polaroid stand almost next door with the early Polaroid Palette, the door of opportunity was opened up and I stepped boldly inside. At that time, Polaroid were best known for their "instant prints" which were a fantastic innovation, especially for parties. This would be the equivalent of the ability to take a snap with your phone and post it directly to Facebook for all your mates to laugh at. The Polaroid Instamatic was a great hit at parties. You just took a snap with the special camera, pulled out the print and waited one minute for it to develop before pulling off and disposing the messy chemical backing side and wafting the still wet and barely developed print until it was dry enough to pass round your friends.

The Polaroid Palette slide making system connected to your monitor socket on the PC and your computer monitor was then plugged into the back of the Palette so that what you saw on your screen was also displayed inside the Palette box, and a 35mm camera attached to the front of the Palette then took a photograph of that image. The software supplied with the kit controlled the shutter of the camera for exactly the right imaging time. As well as the 35mm camera, there was also a Polaroid Instamatic camera that could make instant colour prints. The

Polaroid Palette Slide Maker circa 1985

35mm camera could take Kodak Ektachrome film and there was even an "instant 35mm slide" film that you could process yourself to make your own slides. This instant film looked like the traditional film cassettes used in ordinary 35mm camera but came with a separate square box of processing chemicals that you fitted your exposed film into and wound a handle, waited for the processing time necessary, and then rewound the handle to deliver the developed film strip which you had to cut and mount using a slide mounting kit.

I began to make my own 35mm slides, mostly using the higher quality Ektachrome 100 film which needed to be taken to a chemists or processing facility to be developed and mounted. I could see the commercial potential in offering this as a service to my corporate customers and so began to promote both the design of business presentations and production of 35mm slides to my customers and prospects. I remember very clearly my first order from a client which was for some basic text slides with white text on a black background. Execuvision had a very limited range of fonts but it was more comprehensive than the single Courier font which text applications used. This first customer was, very ironically given the quality of the images, associated with professional typesetting and was thus used to much better results than we could ever hope to offer with this most primitive of early imaging systems. Because it was a rush job I had to use the Polaroid instant 35mm slide film and can clearly recall processing this film in the kitchen of my Oadby home and as carefully as possible hand mounting the slides.

Mass Mitec's award winning stand at the Leicester Exhibition Centre with Sue Berry, myself and Caroline Austin

In 1985 we "pushed the boat out" to order an exhibition stand at a Leicester Trade and Commerce show in the now defunct Exhibition Centre not far from where the Leicester Curve theatre now stands. There was only myself and my first employee Sue Berry working at Mass Mitec at the time so I co-opted one of my Rentacrowd friends Carolyn Austin to help out. We all wore special outfits which included straw boater hats and red bow ties and, as well as kick starting this new venture into presentation graphics, won us first prize for the best stand and a cut glass fruit bowl that I still use to this day. Although Execuvision worked very well for us and was upgraded with a newer version with much more clipart images, it was still very limited in the quality of its graphics and the results that could be obtained using the Polaroid Palette. All of the clipart was in a "raster" or bitmap format meaning that each image was made of individual colour pixels that would deteriorate in quality once they were magnified unlike the familiar vector images made of drawn coloured polygon shapes which can be magnified to any level without degrading quality.

We saw the first improvement in presentation quality with the release of a new software package that removed many of the previous limitations of Execuvision. 35mm Express had been specially written to get the best from computer graphics and imaging systems like the Polaroid Palette. It had far better quality clipart and fonts and the ease with which you could design many different types of slides was a big improvement. The technique which 35mm Express software used to control the camera on the front of the film recorder also effectively doubled the resolution of the slides to around 640 x 400 – still considerably less than a 1 megapixel digital camera would be able to achieve some 15 years later. Once the popularity of this new strand of Mass Mitec business became established, it was clear that my corporate clients were far more interested in how they could either use our design and output services or buy the equipment and software themselves to use in-house than they were in the strategic consultancy aspect that I had originally envisaged would be our core business. So it was in 1965 that Mass Mitec Presentation Services was born and I moved out of the Venture Business Forms offices to take out a 3 year lease in the ex-hairdresser business at 52B London Road Oadby, just above Chris Sharpe Motor Bikes.

In 1985 I moved into a tiny two roomed cottage in Meadow View, Oadby, just fifty yards from the front door of Mass Mitec's new office base at 52B London Road. It was after we made the office move to 52B that I began to recruit the staff who would give Mass Mitec its unique character. I was looking for someone to help me with marketing and I interviewed the young graduate Rachel Smith who would become one of my twin support pillars over the next few years. Rachel was a very shy, attractive and quietly spoken young lady when she first sat across my desk at the job interview. Her family came from the village of Old in Northamptonshire, home of the transport firm "Knights of Old" that you can still see through your train window if you travel on the main line between Kettering and Market Harborough. She had great enthusiasm and showed a lot of interest in our presentation services work. Rachel was to go on to have a massive impact on Mass Mitec before she eventually left some years later.

The other twin pillar was a very unusual and geeky young man called Gary Capewell. I had known his father Ray through my wife's volleyball club activities. Ray was the owner and Managing Director of a printing firm and he had already done some print jobs for me on such things as business cards. It was whilst I had my office at Venture Business Forms that Ray had first mentioned his son. On a social occasion at the volleyball club he asked me for some advice on what home computer to buy Gary for Xmas. This was one of those occasions which are the exact reverse of "asking the advice of the mate in the pub". Ray assumed that because I had worked for IBM that I must know all about home computers as well. Gary had indicated his desire for the new Sinclair ZX81 computer that plugged into your television and could be programmed in the Basic language. Not wanting to show my lack of knowledge, I suggested that it would be a good purchase as they were very popular in those days.

Ray Capewell seemed satisfied with this advice and it came of something of a surprise when he challenged me in the street a few months later when we bumped into each other. He said "You told me to buy Gary that ZX81. I've just found out that he's been writing computer games in his bedroom and selling them to W H Smith Stationers. He's been getting cheques in the post and because he is too young to have a bank account, the bank needed my involvement to cash them." Apparently Gary had, like the Oliver twins who founded Blitz Games around the same time, decided that he could do a better job than what was available in the shops and had started designing his own. He would build a game and then video himself playing the different levels before sending the video cassette off to the buyers at W H Smith and getting the order.

With this talent, it wasn't a difficult decision for me when Ray asked me if I had any jobs that might suit Gary and I agreed to interview him. The interview was one of the most bizarre interviews I have ever been

Sinclair ZX81 home computer

involved in. Ray brought Gary with him to the Mass Mitec offices in Oadby and proceeded to not only sit with Gary at the interview but also answer most of the questions I asked Gary, who came over as sullen with an unsmiling hangdog expression on his gaunt face. Knowing of his abilities however, I offered him a job and he, with Rachel, became the backbone around which Mass Mitec would grow and prosper. Gary taught himself the graphics software we were using at the time and everything involved in designing slides and producing the output. Rachel was not only an innovative and capable marketing person but she also had a special way of handling Gary and they became a powerful and valuable team. Rachel also proved herself very good at winning new business and gained some important new clients, including Pannell Kerr Forster Accountants in Nottingham.

I had now settled in to my new home in Meadow View with my pet cat Bertie. My home, which I bought outright for £12,000 to avoid paying a mortgage so I could plough everything back into the business, had one living area downstairs with a raised kitchen space and some tiny stairs leading to the bedroom and a curtained off section with a toilet and shower. There was very limited storage space but that was no problem because my fashion collection of clothes was a fraction of what it became in later years with the advent of cheap clothes from Primark and Marks and Spencer. I was not dating anyone at the time and enjoying fellowship of the Wigston Church where there was a growing band of new recruits from Kibworth Drama Group. I was working very hard and sometimes long hours, and I remember on one occasion starting work in the early hours of Monday morning and not stopping until late the Friday afternoon of the same week. I had intended to go out that Friday evening and planned to have a couple of hours snooze on my downstairs settee only to wake up 12 hours later still fully dressed!!

Our staff numbers grew, largely through a Youth Opportunities (YOP) scheme that was being administered by Leicester ITEC and which offered young people subsidised placements with local companies that gave the youngsters work experience without adding too much to the salary bill of the small companies the scheme was designed to benefit. Sue Berry eventually left to start a family and she was replaced by Angie Wells from Lubenham. The two YOP recruits we gained during the 1985

– 1989 period were Kevin Brooks and Sonia Hanspal and Mass Mitec began to feel a bit like a "Crazy Gang" of real characters who generated legions of stories.

Kevin lived in nearby Wigston and was always very subdued whilst I was in the office, apart from his legs and feet which he used to move up and down in nervous agitation. He mainly helped Gary and worked under his supervision doing some of the graphics work. I discovered that Kevin had never really travelled anywhere even though he was already seventeen. He had never been to London and what he knew about the capital city was only what he had seen on the TV or in films. I was quite staggered to find this out and so when there was a graphics exhibition at Olympia, I took him down on the train and he was like a kid in a candy store with eyes wide open at the red London buses and the tube train we took to Kensington. Kevin only demonstrated his extrovert hidden personality and humour when I was safely out of the office and his favourite form of entertaining the staff was "pogle diving". This took the form of filling one of the large cardboard boxes in our store room with polystyrene packaging chips and bubble wrap and diving into the box from the table. The spirit in the office was so good and the team worked so hard that I turned a blind eye to these excesses.

Sonia Hanspal was a well-built stocky young Asian girl who came from a very traditional family whose expectation for Sonia would be an arranged marriage. Sonia was very loyal to her family and their traditions but she did fancy the man who worked in the motor bike shop downstairs and would use any opportunity to catch his eye. She invited the Mass Mitec team to an Asian celebration of some kind one evening in the centre of Leicester and we were quite honoured to be the only non-Asian guests. We were made very welcome by all the Asian people there and Sonia did her best to translate the words of the traditional Indian songs that the band played.

Our success in building a strong brand in presentation services and good partnerships with both graphics software developers and distribution companies meant that we were able to travel to industry exhibitions not just in the UK but also in the USA. One of my first business visits to the USA was to the Siggraph exhibition in New York. At that time I had just changed my drama group and had left Kibworth to join the larger Market Harborough Drama Society (MHDS) who had their own small but splendid 100 seat theatre. One of the leading lights of this group was John Bass, a frequent visitor to the "The Big Apple" on business, and he gave me some useful tips for survival as a tourist in the area around 42nd Street where I had a hotel booked. He suggested to always walk purposefully at the edge of the sidewalk as if you know exactly where you are going and don't stop to look at any maps. He painted a picture of New York at night with dark shop doorways where unsavoury people lurked in wait for unsuspecting tourists. New York was as exciting as I had anticipated, with iconic jets of steam rising from the street, and as I lay in bed at night I could hear the familiar police and fire truck sirens endlessly crossing the streets below me.

The Siggraph exhibition was my first exposure to Corel software and a functional drawing package. I recall that it was their new release of Corel Draw 2.0 that was being demonstrated and I bought a copy for our office use. Corel Draw became one of our most important packages for customised graphic design, especially for creating company logos. As it was my first visit to New York, I treated myself to a show on 42nd Street and saw "Me and My Girl" starring Jim Dale who was most famous for his gormless characters in the "Carry On" films of the day. It was this show that made me realise just how talented Jim Dale was and how his film typecasting belied his ability to act, sing and dance. I felt very proud that he was a "Limey" who I believe originated not far from Leicester in Wellingborough, Northamptonshire. I also visited Grand Central Station to see the platform for the famous Chatanooga Choo Choo of hit song fame and even had my shoes shined there as a tribute to the musical lyrics. The best experience of all however had to be the helicopter flight around New York which I took as sunset was approaching. This ride gave me the best photographs I had ever taken as we flew round the Chrysler building and over the deceptively diminutive Statue of Liberty.

The Siggraph conferences/exhibitions became an important part of my strategy for keeping up to date with the latest technologies and I tried to give my staff the same opportunities by sometimes taking them with me on these trips. On one occasion I took Gary Capewell with me and we shared a twin room in a California Hotel near the conference which was held at Anaheim. The conference location gave us both a chance to see Hollywood and provided me with my first of many subsequent trips to Disneyland. We took in Hollywood, the Universal Studios and Knotts Berry Landing. I adored the theme Parks and have distinctive memories of a ride called Montezuma's Revenge which certainly lived up to its name. In the Universal Studios we had a tour of the film making lots which included audience participation. I volunteered myself at the 2001 Space Odyssey lot and found myself chosen to play the role of a Russian Astronaut in a space rescue. I was fitted out in the space suit and suspended on wires in front of a huge blue screen. Acting on instructions I could hear from a speaker in my spacesuit I floated across this backdrop to couple with an American astronaut to simulate the rescue. This scenario was used to demonstrate the "blue screen" (later to become green screen) techniques used to create the special effects (SFX) used in movies of that time. I was later to meet up with someone who had worked with this technology in Hollywood on the making of Superman 1. When it was time for Gary and I to leave, I had a shock bill for the hotel phone. Gary had been calling a new American girl friend he had got close to and had run up over $400 dollars in call charges which I later had to deduct from his salary.

Our regular staff education outings were at the annual "Softeach" events organised by the Distributor Softsel from whom we purchased most of our software for resale. It was one such workshop run by Software Publishing (SPC) that led to Rachel Smith winning a trip to San Francisco. Over the period we were based at 52B London Road, our turnover grew rapidly to around £250,000 per annum. As the hardware and software technologies got better, we kept one step ahead of our competitors and our clients. The graphics capabilities of the desktop PC began to improve rapidly and the original 4 colour limitation of the first IBM PCs grew to 16 colours and then 256 colours. The resolution also improved from the original CGA of 320 x 200 pixels to the VGA mode of 640 x 480. Our 35mm Express software had enabled us to get the very best out of our graphics for the business presentations we were designing and not only offered better quality and sharper slides but also provided a palette of 64 colours at a time when most people could only get 16 colours.

In these days at 52B, the IBM PC still had its DOS operating system and required text commands to load programmes and perform other housekeeping tasks. The Windows operating system with its graphical user interface was still in the future. We began to buy and resell new graphics software as it came onto the market, quite often launched at the Softeach events held near Heathrow airport at the Renaissance Hotel. Harvard Graphics became the new kid on the block for designing business presentations and it was this software that was the beginning of mainstream presentation graphics that managers in our corporate clients could use for presentations. In these early days there were no colour inkjet printers and the only desktop printers were of the dot matrix type or later, the early desktop laser printers. The dot matrix printers could only manage cartoon type images and were wholly unsuited to business presentations.

Our quest to continuously upgrade our technology led us to buy hardware like the Quintar 1080 processor box which effectively improved the text quality of 35mm slides. I believe that Mass Mitec was probably the only company in the UK using this technology which was designed to overcome the limitations of the PC graphics cards by providing an interface box between the PC and the Polaroid Palette film recorder. This interface had plug-in font cartridges and a separate higher quality RGB monitor to preview the slide designs. The Quintar 1080 improved still further the competitive advantages of 35mm Express. As well as Harvard Graphics software from SPC we also added Lotus Freelance Graphics and a new product from Microsoft called Powerpoint. For special creative designs which were beyond the scope of these end-user focussed packages we used Corel Draw, especially for designing client logos.

It was very early on in Mass Mitec's existence that we became involved in data projectors for presentations. The projectors available at the time to connect to the IBM PCs which offered colour graphics capabilities had 3 colour lenses and were very heavy. I used one of these projectors at a seminar I ran in my old IBM offices in Nottingham. It was a "Barco" projector manufactured by the industry leader at the time, Barcovision. This massive projector sat on a trolley in front of the projection screen and required about half an hour to align its 3 projector lenses to get a clearly focused image. This process involved projecting a test image of a grid onto the screen and adjusting the projector and its position to get a clear focus for each lens before you could project anything. This delicate process meant that if anyone accidently brushed against the projector trolley, the presentation was effectively ruined. At this seminar on presentation technology was what came to be one of our most important future clients, Fisons Pharmaceuticals, who were a massive organisation based in Loughborough. At the end of my presentation their representative asked if I could help them out with a corporate presentation they needed to do in Lisbon, Portugal. They told me they were short of staff and wondered if I was free to do all the set up and organising for them. They agreed to cover all my costs and pay my normal daily rate which, at the time, was £250 per day.

The arrangement was that I would take their presentation slides on a Compaq portable computer with me on the plane and make all the arrangements to set the computer and a locally hired data projector up in Lisbon before bringing the computer back with me. They booked the flight and also arranged a first class hotel in Lisbon. When I arrived at Lisbon airport and was standing in the queue waiting to go through passport control with my weighty luggable Compaq that so resembled a sewing machine case, I felt a tap on my shoulder and a fellow passenger on the flight introduced himself as a Portugese man living in Lisbon. He was just coming back from London after attending a two week course on computer graphics and he wondered what I was carrying. I explained that I had been commissioned to bring this Compaq portable to Lisbon

Compaq 'portable' computer

and to set it up with a data projector at an international conference. This happy coincidence led to him offering his help in all these tasks because he was a native speaker and also knew about graphics. In the event, he was so hospitable that he effectively did all of the work for me out of the goodness of his heart and even arranged to take me out one evening with a former girl friend of his for company. I quickly thought that if this is what running your own business is like, bring it on!! The only downside of his hospitality came when he took me out for lunch at one of the best fish restaurants in Lisbon and our meal over-ran leaving us barely enough time to get to the airport. He drove like a madman to get me there and, when I checked in the airline, upgraded me free of charge to First Class.

I do not remember the exact circumstances now, but I was later given or bought very cheaply the Compaq portable and we acquired our own data projector, a Thompson three lens machine similar to the Barco but with its own curved silver coated screen that was infinitely easier to set up. We began to use this in our own seminars, many of which were held in the local Moat House hotel in Oadby. To market these events we would use mailshots by both post and telex. From a mailshot to about 600 people we would often get as many as 100 actual attendees, a remarkable return compared to what would be achievable today. The use of our own data projector for these events was also the beginning of our involvement in reselling data projectors which would later become a growing part of our business.

As well as 35mm slides, overhead transparency "foils" were still an important presentation aid at that time, mainly because 35mm slide projectors required some organising and management whereas an overhead projector using foil transparencies was much easier, and the presenter could easily change his slides by replacing one transparency with another. It was for this reason that we looked for a colour printing solution that could produce good quality overhead transparencies and prints. Our quest for such a printer led us to the start of a long relationship with Calcomp who were the industry leaders in plotters for large scale engineering drawings. Colour plotters at that time used coloured pens to draw diagrams and plans in colour. We did acquire a small desktop plotter ourselves but these devices were very limited in what they could produce and our own plotter only had four pens.

Calcomp four colour imaging ribbon

Calcomp had a desktop colour printer which used wax thermal technology to produce full colour transparencies and prints. The process involved overlaying four colours and using heat to bind the colours to paper or transparency to produce a good quality dry output in 2-3 minutes per print. There were software drivers available for all the software packages we were using and this gave us an opportunity to drive even more sales for both our presentation design and our hardware and software sales. We also acquired an HP 3000 laserjet printer which significantly improved the quality of our mailshots and in-house printing capabilities and we began using a desktop publishing software package called Ventura Publisher.

This was a very happy period in my life. I had a good social life with my drama and with the Wigston Church and I had a happy team. Our neighbours on the opposite corner of Meadow View were "Radio and TV Services", a television and radio repair business in the days when these services were still necessary. The middle aged husband and wife who ran the business were very friendly and gave me a kitten to keep my ageing moggy pet Bertie company. Bertie was a black and white slim animal who was always very affectionate to me and liked to rub foreheads with me. This new kitten was a tortoiseshell female who was plump but very shy of people. Bertie accepted this newcomer to the household very well – he took most things in his stride and Tiger came to live with us. I would see nothing of Tiger all day but at night time I would hear the cat flap bang and the thump of her paws as she raced up the stairs to jump on the bed and snuggle down on top of my chest for the night. Sadly, one day when I was doing some training in my offices, a passer-by came in to tell me a cat had been knocked down. It was Tiger

Montage film recorder

and she was lying dead but apparently unmarked on the pavement. I carried her into my house and laid her on the bedroom floor until my day's work was done. The client being trained was very understanding and offered to re-arrange but we struggled on and later that day I buried Tiger amongst the roses that originally came from my grandmother's home in Boston.

After Mass Mitec moved into its more prestigious address at the White House, we continued to strive to provide improvements in the quality of our services and to establish ourselves as one of the UK's leading specialists in presentation and business communications technologies. The graphics capabilities of the IBM PC continued to improve and more powerful graphics cards became available with more colours and higher resolutions. In the early PC colour graphics computers, a lot of the graphics processing was done by the main computer, but the newer graphics cards took away a lot of the processing work from the PC and made the improvements that resulted in the development of the rich 3D graphics used in games and virtual worlds today. The screen resolution improved from CGA (320 x 200) to VGA (640 x 480) and to SVGA (800 x 600) and the number of colours in the colour palette that could be displayed on the screen increased from 4 to 16 to 256 to 64,000 and with these improvements came much more realistic and photo like images. One of the major issues from the beginning of the presentation slide business was the quality of text which, when a 35mm slide was magnified on a large screen, would show ragged edges instead of smooth curves. This was because the graphics cards in those days had very limited, if any, anti aliasing capabilities. To overcome this we had been using the Quintar 1080 but we took another step forward with a Montage film recorder that operated a bit like the Polaroid Palette but had special "font wheels" inside the body of the Montage. These wheels looked like a disk with the type characters as holes in the wheel and the driver for the Montage would spin the disk to the correct position for the light to go through the hole and form much better characters on the slide.

Like many of the products we were acquiring to both use in the business and resell, the Montage was a USA product which we had to arrange special shipment for as there were no UK agents or distributors. I bought our Montage film recorders from a firm in Boston Massachusetts managed by a man called Ray Stafford. I got to know Ray quite well through our dealings over the telephone and he told me that he was about to get married and that he and his bride would be coming to the UK for their honeymoon, starting in London and then travelling up through England to visit his namesake town of Stafford before going on to Holyhead to cross to Ireland. I invited him and his new bride to stay the night at the White House and arranged that Carol and I would drive them up to Stafford before going on to Ruthin Castle where we had had our own honeymoon. The idea of staying in an ancient UK castle really appealed to them both and so we made the booking at Ruthin Castle for them and I told the hotel that it would be their honeymoon. I asked if it would be possible for Ray and his bride to be made Baron and Baroness for the mediaeval banquet we had booked the same evening. It was a fantastic experience for these two Americans to be honoured at the top table of a proper mediaeval banquet in a genuine piece of British history.

Mass Mitec had established itself as one of the leading presentation graphics companies in the UK with an impressive list of blue chip clients and a number of awards. We were awarded Northern Dealer of the Year by Software Publishing (SPC) for our sales of Harvard Graphics. We were the very first UK company to be accredited by Microsoft as a Powerpoint Centre of Excellence and we were also accredited with authorisation from Corel and Lotus Freelance. SPC ran a dealer competition through our distributors Softsel that was designed like a grand prix race with three laps. Each Harvard Graphics dealer was assigned a sales quota based on their size and turnover and every month the sales orders were compared to these quotas and the best performing dealer was awarded a prize. The prize for the winner of each of the first two laps was £1,000, which we won as the best performers for laps one and two. About three weeks into the final month, we were still ahead on sales, so SPC arranged for the prize to be used for publicity at our offices. It was a Panther Kalista sports car worth £10,000 and a fantastic incentive. I learnt later that Mass Mitec had been in the lead until the very last day of the competition when the SPC Marketing Manager tipped off a rival company with information on what they needed to order to win. I understand that this Marketing Manager then left SPC later to join the winning company leaving us with the consolation of second prize which was the use of a Ferrari Sports Car for two days. Even though I am not a petrol head, this car was a real head turner and I used it as much as possible to give my friends and family a spin, even taking it over to Boston to show off to my parents.

It was at Softeach that we first saw and obtained our first copy of Windows 3.1 as the new graphical user interface to help Microsoft compete with the Apple computers that were gaining in popularity. It seems incredible now that we had established ourselves as a leading graphics company without even the benefit of the computer mouse that was now being introduced to support Windows. The early version of Windows was quite poor and riddled with problems. We continued to use our DOS packages but these were gradually phased out as Windows eventually took over as the main operating system. Once Rachel and Gary had left, Carol's youngest son, Christopher joined our ranks as a trainee graphics designer alongside his brother Paul who had by then established himself as a good graphics designer and, despite his lack of further education, he had a good manner and responsible attitude. Christopher also took to computers very well and later went on to get some good jobs in mainstream computing in the banking industry.

At its peak Mass Mitec had ten employees and our portfolio included presentation design, bureau imaging, hardware and software sales and presentation software training. When the White House offices had been converted from the original barn we had incorporated a dark room and a 35mm slide film processor to give us complete control over the process of making professional quality visual aids. We had developed good relationships with our main distributors and manufacturers like Polaroid and Calcomp and we were now entering the age of digital imaging and the dawn of many of the technologies we take for granted today.

Canon Ion Still Video Camera

It was around 1990 that we acquired our first electronic camera, the forerunner of today's digital cameras. It was actually technically an analogue camera that saved still images on small disks in what today would be a poor resolution of about 320 pixels by 200. This camera had a video output lead that could be plugged into either a TV or a video capture card on a PC. This was very primitive and in the very early days of tools to capture and manipulate video on a desktop computer. We had acquired video capture and video output cards from Targa for our own experimentation and also began to explore the use of early animation software such as Autodesk Animator. It was this albeit limited experience that got us an early commission with Rolls Royce Aero engines in Derby. They had seen the potential of computer graphics and animation for their training activities and had developed an animation to illustrate the procedures involved in replacing a faulty aero engine in a location without permanent maintenance facilities. Their animation showed the plane landing and jig assembled under the wing holding the faulty engine and eventually the replacement engine being wheeled into it place before seeing the plane take off. It was pushing the boundaries of the animation software but we were able to deliver the video recorded on VHS tape to them.

In 1992 Kodak introduced a new technology called Photo CD which significantly enhanced our ability to be able to digitise photographs for use in presentations. There were very early desktop scanners in those days and we had already acquired a flatbed scanner capable of scanning up to 2400 pixels per inch but the Photo CD service offered far better results. In these days before any kind of viable digital camera, this service was a major improvement. I first saw it advertised in a national newspaper which announced that the first place to offer this in the UK would be Selfridge's store in Oxford Street in London. I wasted no time in taking some photos with my 35mm camera and catching the train with my undeveloped film cassette in my pocket. The turnaround time was about a week and I went to collect our very first Photo CD disk the following week. I was delighted by the results. All our computers had CD-ROM drives by this time and this new service offered many possibilities. It wasn't long before the availability of the service reached Leicester where we were then able to go to the Jessops retail outlet on Hinckley Road.

Our use of video capture technology led to an amusing incident in the early 1990's. I had arranged to do a demonstration of video capture to the Forward Trust Group, a financial services company based in Birmingham. To show how easy it was to capture video, I recorded that morning's television news onto a videotape that was left in the video player in our lounge. I took the player and the tape into the office and plugged it into the PC and began to fast rewind the tape to the start of this news bulletin. As I was rewinding and watching the video flash past on the screen I noticed that the video that had been in the machine was a soft porn video. I was angry that this could have led to a lot of embarrassment at my demonstration to this important client and so Stephen, who at the time was around sixteen, was admonished and promptly left home in disgust. It later turned out that this video had been used by the boys in the family as their sex education tool and had been handed down through the family as each boy grew bored with watching it. I learnt later that it had been the youngest boy Christopher who had

Baby Amie Blake captured on the Canon Ion camera and printed on a Laserjet printer in 1990

left it in the machine. Stephen did return home after staying at his mate's house for a couple nights and he went on to great things in the RAF and later at University where he got a first class honours and a post-graduate doctorate.

After Rachel and Gary left to join a semi-competitor who specialised in organising conference audio visual staging and graphics, we were left with a core team which now included a graphics designer called Brett Hodges, my two stepsons Paul and Christopher, and Angie Wells who looked after the administration and accounts. Angie lived with her parents nearby in Lubenham and was quiet and reserved but had a good sense of humour. They would be later joined by Ruth Kanyeihamba, a graduate recruit supported by another Government scheme. The consequences of Rachel and Gary's departure were that the business suddenly became highly dependent on me, not only to win new business, but also to direct the staff and be a revenue earner myself.

The main problem for the business was that none of the staff had Rachel and Gary's combination of experience and innovation, and since our fixed overheads for running the business were very high by this time, it placed a large burden on me in a situation where I could not afford to employ the kind of experienced staff that would resolve this issue. What the business needed was a regular and almost guaranteed income stream that could make Mass Mitec a smooth running "factory operation" that did not need constant attention to bring in new business. The technology that created this opportunity came in the form of the Diskfax invention described in the telecommunications chapter. This device had the potential to overcome time and distance and enable us to offer professional presentation graphics with a fast turnaround almost anywhere in the UK. I decided that what was required was a business partner with a large number of High Street outlets and a motivation to sell our services to their customers. I identified High Street Print/Copy chains as the best partner and targeted the one with the largest number of outlets, Prontaprint.

The Prontaprint Head Office was based in Darlington up in the North East of England. Darlington was at the heart of the transport revolution as the very first railway built was the Stockton and Darlington where Stephenson's famous "Rocket" locomotive ran. Prontaprint ran a franchise operation across over 200 stores in the UK with one or two international branches. Their main core business at each High Street retail outlet was printing of brochures and leaflets and general photocopying. Many of these franchises were small family run outlets and Prontaprint HQ allowed the franchisees to choose from a portfolio of services according to their preferences and local knowledge. My pitch to Prontaprint was based on the quality and margins of the additional business I could bring the franchisees if they offered 35mm slides and Overhead Transparencies from client presentations to their portfolio. The use of a Diskfax terminal meant that, unless the franchisee wanted to get involved in designing presentations themselves, there would be no need to have an IBM PC on the premises. It was mainly corporate clients who were using programmes like Harvard Graphics and Powerpoint to design presentations, so the quality of customer who would be likely to use the service would be of a higher standard than most franchisees would see in their stores.

It was a very emotional day for me when I visited Darlington and heard Prontaprint management tell me that they had accepted my proposal for Mass Mitec to act as their National Presentation Centre. I had made up a Franchisee Pack with marketing materials including sample 35mm slides, Overhead Transparencies and Colour Prints, along with some instructions on how to sell the services and use the Diskfax. The start-up package I would offer each franchisee for a fixed price of £1500 was a Diskfax unit, Harvard Graphics software, the marketing pack and an on-site half day workshop to act as a training session and a launch event to their local customers. I also agreed to pass on any names of existing customers in the franchisee area.

Following the Prontaprint Board agreement they arranged for me to make a keynote presentation about the new service at their annual conference and to have a stand in the exhibition area. The Prontaprint annual conference was a major event for all the franchisees and was very well attended. It was also a big social and networking event that franchisees would bring their families to with big name entertainment of the likes of Bob Monkhouse (who was tremendously professional) and Willie Russell. At that first national conference to launch the National Presentation Network I signed up 70 franchisees over the weekend

Solitaire digital film recorder

which, with the good deals I had done on Diskfax and Harvard Graphics (normal RRP together of around £1500) left me with a margin of about £500 to cover my workshops.

Almost immediately after the Prontaprint conference I began a national roadshow of franchisee launch events from one end of the country to the other and even across to Dublin in Eire and Belfast in Northern Ireland. We had bought a Solitaire film recorder capable of making slides at a very professional 4,000 lines resolution and all of my staff had been trained how to process 35mm slide film in our darkroom. The Prontaprint deal earned us a lot of great publicity and within about 3 months we were seeing around £10,000 per month of slide business. The franchisees would charge a flat rate of £5 per 35mm slide and we would charge the franchisee £3 plus any post and packaging. Their customers would simply bring in the Powerpoint or Harvard Graphics presentation on a 1.2MB floppy disk (in the days when presentations were a much smaller file size) and the franchisee would place the disk in Diskfax and dial our office Diskfax number. As soon as we received the file (typically less than 5 minutes transfer time) my staff would complete an order form and review the disk to check the presentation. The franchisees would also normally send a fax confirmation at the same time.

So it was that the Mass Mitec office became like a presentation services factory that I did not need to supervise on a daily basis. In the first six months of operation I travelled thousands of miles running these workshops and training sessions with many memorable incidents. We guaranteed a 48hour turnaround time but the normal turnaround time was 24 hours from the time we received the disk file to when the set of slides would arrive at the franchisee anywhere in the UK. This meant that we would process the film, mount the 35mm slides, package the order and take it directly to our local sorting office in Market Harborough before 18:30 in the evening. For more urgent cases when the order was not completed in time to catch this deadline, we used the services of a local courier who offered us special rates. There were many memorable incidents including one occasion when we got a last minute order from Glasgow that was very important to the franchisee. We had missed the deadline so at 10pm at night I got in my car and drove up to Glasgow myself to be waiting at the franchisees shop when he arrived to open up. This attention and commitment to good customer service generated a lot of goodwill amongst the franchise network.

Once this operation was running smoothly, it allowed me to concentrate on building up the corporate business including graphics training courses, roadshows and the sale of data projection equipment. The deal with Prontaprint had a strategic intent behind it because I was aware that the 35mm slide business had a limited shelf life and that companies would increasingly be using data projectors for their presentations and also that overhead transparencies would disappear forever. My plan was to help Prontaprint establish itself as the High Street presentation services specialists and that my 35mm slide business would evolve into the sale of different kinds of presentation technology through the Prontaprint retail outlets. As it turned out, this strategy was flawed for a number of different reasons but ultimately because Prontaprint were taken over by a large printing concern whose main priority was feeding their print factory through a chain of retail outlets – the same business model I had provided Prontaprint.

The Prontaprint partnership did serve Mass Mitec well not only for the volume of business it brought in but also because of the customers it attracted who wanted to do more complex presentation business directly with me. Amongst these customers was the Atomic Energy Authority whose headquarters were in Harwell Oxfordshire. They commissioned me to do some Harvard Graphics training at their Doonreay power station in the far north of Scotland. This meant me travelling to Manchester airport to climb on board their private 9 seater light aircraft one afternoon to be flown to Scotland for an overnight stay in a quaint local town before training some staff at the offices on the power station site. At the end of the training around 4pm, I joined a small group of people walking out of their main reception across the tarmac of the airfield to be flown back to Manchester. It was just like catching a bus and I was seated directly behind the pilot sharing a great view with him through the cockpit window and handing him periodic cups of coffee. He warned us all that some bad weather was forecast to come in from the Atlantic so we had to climb high above the storm. By the time we were approaching Manchester, the clouds enveloping the plane were pitch black with rain hammering the windscreen and zero visibility. My mind took me to the comedy film "Airplane" starring Lloyd Bridges and Leslie Neilson and I half expected a river of sweat to come pouring from the pilot's head. The buffeting of the plane seemed to last an eternity as we descended lower and lower until suddenly we broke through the clouds and I saw to my relief the runway about 200 feet below us with the commercial airlines queuing on the tarmac waiting for our landing to clear them for take off.

Another very important customer I acquired during this time was Thorntons plc who manufacture my mum's favourite chocolate and have their head office near Alfreton just up the M1 from Leicester. In my time working with Prontaprint I had also run special roadshows to which the franchisees could invite their customers to be shown good presentation practices. The roadshows were not overt sales pitches but were meant to be educational and supportive of the franchisees. I think it was through one of these roadshows that I made the connection with Thorntons who were planning some major changes to their business strategy which included new styles of shops. I was able to offer them a one-stop shop for a programme of activities that included presentation design and advice to their board, production of professional quality audio visual aids and the running of roadshow events to launch the strategy. I thoroughly enjoyed my time working with Thorntons and had the privilege of meeting the family who started the business in the early part of the 20th century by making their famous toffee in the back kitchen in Sheffield.

I worked with the Thorntons board members for several years and it was probably because of the length of this relationship that they were so understanding about an unfortunate incident at their annual shareholder meeting one year at the Derby Playhouse. It was at the time when their adverts featuring the song "Heaven is a place on Earth" were being run and they wanted to use a mixture of 35mm slides and video for their corporate presentation. I had set up the slide and video back projection system on stage behind the main screen and, about a minute into the video presentation, I realised that I was watching the video the correct way round, which meant that the audience had a mirror image view. My hopes that no-one had noticed were scotched when the very first question asked by one of the shareholders was "Has Thorntons now changed its name to "Snot n Roht". I could have died with embarrassment.

It was also through the Prontaprint National Presentation Network that I met a lady who was to become my most valued customer and with whom I have worked for many years and become good friends. It was very common for the franchisees to ring our office for advice when a customer had a request they didn't have the knowledge to deal with. I got a call from Prontaprint Hemel Hempstead one day saying that they had a lady asking about how to make presentations with an Amiga computer and could we help her. The franchisee gave me the organisation's name, Foundation for Water Research based in Marlow just off the M40, and their Chief Executive Caryll Stephen's phone number. It was an odd request but I rang Caryll and offered to visit her offices to discuss her requirements. She explained that she had an Amiga computer at home with graphics capabilities and she wanted to make her own presentations. It was this meeting when I explained how most business presentation were made on IBM PC hardware using standard packages such as Harvard Graphics and Powerpoint and this led to the first of many orders for a whole variety of services including setting up and looking after her home office technology.

By this time, Windows had established itself and although Software Publishing had introduced a Windows version of Harvard Graphics, it soon began to lose ground to Microsoft Powerpoint as the industry standard presentation software package. Lotus Freelance which had been an early competitor completely lost its way and effectively disappeared leaving the way clear for Powerpoint to dominate. Corel Inc had introduced other packages to their stable including a powerful painting package and it was Corel that my graphics staff mainly used to design corporate logos for client presentations. They "digitised" logos by scanning a good quality printed copy of the logo using one of the desktop flatbed scanners that had entered the market and then using Corel Draw to make a vector (drawn shapes) version that would retain its quality however large it was blown up. Digitisation of corporate logos was a skilled business that not only gave us a competitive edge but also an entry into many blue chip companies. The evolution of Microsoft Powerpoint has been a classic example of the theme that runs through this book like a stick of rock. In the early days of all these technologies it had required the skill and experience of a knowledge professional to get the best out of the technology. Now professional looking presentations are being created every day by ordinary untrained people all over the world as mankind's intelligence has become embedded in the software packages that we all use. The phrase "death by Powerpoint" is in common parlance these days to reflect the fact that having nicely laid out slides with professional looking background templates and colour schemes does not ensure that the presenter can communicate effectively to the audience. Instructional design, and the understanding of how to have the maximum desired impact is still the domain of human beings who have the intelligence and sensitivity to understand and influence other human beings.

I worked regularly with Caryll Stephen and FWR in Marlow and acted as their technology test bed so that when I identified a new software or hardware solution and trialled it within my own business they would often follow along the same path. I introduced Caryll to data projectors for her business presentations. The early projectors that had been available during the start up of Mass Mitec had used old RGB cathode ray type technologies similar to those used in Televisions and the early bulky PC monitors. These had been tricky to set up and very bulky. Single lens projectors that were more "plug and play" and that could handle the higher resolutions such as VGA and SVGA were starting to appear although these initial machines had limited brightness and quite saturated colours. This worked well for most business presentations that

InFocus projector using DLP technology

showed text, graphs and a bit of clipart but were poor at properly reproducing colour photographic images.

Texas Instruments developed a projection technology called DLP (Digital Light Processing) that revolutionised the data projection market because of its superior ability to reproduce photographic colours and video. I arranged for Caryll and her husband Derek to visit my audio visual distributors in Milton Keynes. The company was set up and managed by Alan Gell and is now called Multivision and in new ownership. Both Caryll and Derek were and impressed and FWR ordered a projector for their boardroom and also for Caryll to take with her for her international conference presentations. The Marketing Manager for this distribution company specialising in audio visual equipment was Dave Barton and sometime later I had organised a special seminar on data projectors at our White House offices. Dave had agreed to bring four projectors of different technologies and prices with him to Lubenham and demonstrate for me. Purely by chance, on the morning of the seminar, I had rung the distributors on a different matter and had mentioned Dave's name in the phone conversation and was asked if I wanted to speak to him. I said "Isn't he on his way to our offices to run the seminar?" and discovered to my horror that he hadn't written the date in his diary and had already committed himself to another appointment. I spoke to him and after apologising profusely he agreed to get his warehouse to immediately send a courier the 50 miles from Milton Keynes over to our offices with four new boxed up projectors.

Over the next 30 minutes, the seminar delegates began to arrive ready for a 10 am start and, after using copious amount of coffee to fill time, I eventually had to start this seminar on projectors without any of the demonstration equipment. I managed to entertain the delegates sufficiently with a mixture of background information on projectors and amusing anecdotes before the courier finally drew up in our courtyard at about 11:00 am and unloaded the boxed projectors into the seminar room. This potential banana skin of a morning actually turned to our advantage when I was able to take each projector out of its box, plug it into my PC, and demonstrate how easy they were to set up and use.

Over the next few years we began to sell video projectors in reasonable numbers to our corporate clients and also developed our skills in designing the screen based presentations that have now entirely replaced 35mm slides as the preferred medium at all seminars and conferences. It was these skills that led to a relationship developing with a Northampton based conferencing and exhibition company. They would begin to involve Mass Mitec in special projects where their client wanted some unusual audio visual technology to attract customers to their stand. The most memorable of these commissions was also my first introduction to what would be called today a "serious game". Their client was a major pharmaceutical company exhibiting at an international conference in Prague on Epilepsy. Our Northants based customer had conceived a stand which would not only allow their pharmaceutical client to run hourly mini workshops on the stand but also would attract visitors by offering a competition with a nice prize of a digital camera at the end.

They had designed some futuristic looking "pods" which would house desktop PCs, and planned to have about four of these pods running a quiz game which showed images, video or text questions that had multiple choice answers. The visitor would choose an answer which, if correct, brought up a type of card game involving turning two cards over and trying to match pairs of cards. If the two cards were the same, they stayed turned over and once all the cards had been revealed, points were awarded for the number of correct answers and the time taken. The game had been custom programmed by a company in London for the client. I sourced all the technologies including the PCs, projector, cabling and switches so that in the intervals between workshops they could display the screens from any of the pods. I joined the exhibition company in Prague to set up and help man the stand which proved to be a massive success and attracted the largest number of visitors.

It was this "serious game" that gave me the idea to try to develop this as another strand to our business and I began to experiment with one of Corel's new software packages called "Click n Create". This was a piece of software that Corel had acquired by buying out the company who developed it. It was a time when Corel's success with the Draw and Paint programmes had given them grand designs on challenging Microsoft with their own suite of office software to rival Microsoft Office, an initiative that ultimately failed. Because FWR in those days often exhibited at conferences, I designed a game especially for them called the "Water Game". There was nothing serious or educational about this game which had more in common with the Space Invaders or Asteroids arcade games.

This experience at the Prague exhibition did however open my mind to the potential of serious games as an exhibition attraction and I saw the model that had been used in Prague as an opportunity to develop our own generic version of this game in which any client could develop their own customised set of questions and answers and promotional graphics as well as having a "front end" which was able to capture company and contact information as well as tracking the scores of each visitor. Neither I nor any of my staff had the programming skills necessary but I was fortunate to hear about a student placement scheme being offered by De Montfort University where we could take on a student and give him a project assignment. We would make sure all his travel costs were covered and give him a modest weekly wage. The student who was assigned to us was a young Greek lad called Spiros Ioannidis and he proved more than capable of designing what we needed and packaging it into a "run-time" installation programme that we could then market. The only work required for any new client would be helping them to design their multiple choice questions and creating the background graphics that would be revealed when the quiz was successfully completed. In hindsight, it is one of my few regrets that we did not properly commercialise this idea, mainly through lack of resources at the time and other priorities. I do believe that this idea has potential even today and may one day be resurrected by either myself or someone reading this book.

Our imaging resources that had been significantly enhanced by the

Fujix Pictography printer

Canon CLC 700 photocopier

Solitaire film recorder were further improved by the addition of a Fujix Pictography printer. This printer was the equivalent in quality to the Solitaire but printed true photographic A4 quality prints and Overhead Transparencies and was effectively a mini self contained photo studio inside its cabinet. The consumables were quite expensive but it earned us a return on investment and was pretty reliable. Throughout the history of Mass Mitec I have been very fortunate not to have had too many equipment failures that might have jeopardised the business. One exception to this was when our Solitaire film recorder broke down and required a new cathode ray tube at a cost of £6,000. This happened not too long after my father had died and I knew we did not have the cash to replace the tube. This was another of the rare occasions in my life when the situation caused me to sob my heart out. Fortunately we got through all these events.

The Pictography printer helped us to win a major piece of business that enabled us to justify the purchase of a very expensive high capacity colour laser printer, the Canon CLC 700. In our time at the White House in Lubenham we had done very little business with the larger companies in Market Harborough, despite our best efforts. All this changed when Golden Wonder needed to make some high quality presentations to the financial markets to raise funds for their expansion. As well as the crisps they are probably best known for, the company also owned brands like Homepride Sauces and products like "Chicken Tonite". Probably because of our proximity and willingness to work all hours to meet a customer need, we were commissioned to produce a large quality of professional quality printed and bound handouts for the City Analysts. We got the files from Golden Wonder and printed high quality masters on our Pictrography before colour copying them "2 up" onto A3 paper, trimming and ring binding them. For this £70,000 order I worked all night and all day and paid a premium rate to Canon for out of hours call-out services in case of breakdown. We had to order really large quantities of consumables for both Pictrography and Canon and delivered the order personally to the location in Watford where these presentations were being held.

With the pace of technological change, desktop colour printing soon became consumerised, and today anyone with a desktop computer or laptop can achieve the same quality as our Pictrography in a shorter time with colour ink jet printers that you can buy from PC World for under £100, and this reflects the trends that I have seen in all the technologies I have been involved with in my lifetime career. With each new technology, the timescale between its launch and adoption into mainstream use by non-specialists has become shorter and shorter so that now it is highly questionable that any knowledge professional could build a business like Mass Mitec, based, as it was, on being an innovative pioneer "living laboratory" of new technologies. This is also augmented by the fact that even the youngest of today's generation are able to adapt to new technologies intuitively, partly because it is a natural part of their world, and partly because all their previous generations have transferred their intelligence to be embedded in these new devices and applications.

Mass Mitec's involvement in presentation graphics and digital imaging drew to a close in the late 1990s when we became more focused on Community Informatics and the power of the web. I did recruit a couple of staff with web design experience, Sarah Keet and Barry Webber, and also employed Dick van Aken as a Sales Manager to help drive this new business opportunity. In the end, as had happened with Prontaprint, I had put too many eggs in one business basket and paid the ultimate price by losing my business and having to start from scratch again.

Imaging technology hardware and software has evolved immeasurably in the twenty five years since Mass Mitec made its first tentative steps with a Polaroid Palette and 35mm slides processed on a kitchen table. Today it is an everyday part of life in which we take for granted that even our mobile phones have inbuilt cameras that a professional photographer would have found valuable in 1984.

We have got the best from imaging technology through our collective intelligence and senses and now it is the time for technology to get the best out of us.

CHAPTER SIX
Video Killed the Radio Star

This chapter is not just about the developments in the technologies which have entertained me and countless millions of other people all over the world both inside and outside of our homes, it is also a narrative about my own passion for entertaining people on stage and on screen, and a selection of the many anecdotal pleasures this passion has blessed me with from the early days of the Boston Gang Show to the later years of the Little Theatre and some limited TV extra work.

In my early years, my home entertainment system was purely mechanical. A wind-up gramophone (is this word still in use in the English language ?) record player with a stylus vibrating in the grooves of a heavy Bakelite 78 rpm record and mechanically amplified to reproduce sound through a big brass horn was the medium that helped to make Elvis Presley and the early Rock and Roll stars famous. It is

Guess who is Miss Suzanne in this sketch from the Boston Gang in 1961?

almost unbelievable that some of the rock classics that are still played with great passion and nostalgia today started life on an old gramophone recording. My grandmother's gramophone with its big brass horn and heavy stylus arm held a great fascination to me as a small boy in the early 1950s. I can still picture her collection of records in their worn khaki sleeves, the maroon coloured central record label of "His Master's Voice Recordings" and the black record label of the "Parlophone" records. His Master's Voice is one of those rare companies that are still with us today in the shape of the HMV record shops we see in our major cities.

My favourite record was the "Jones's Laughing Record" aka "Flight of the Bumble Bee" by Spike Jones and the City Slickers. No matter how times I played this record I always ended up laughing infectiously along with his band, an anarchic forerunner of the Bonzo Dog Doodah Band so popular in my late teens. Novelty records were quite in fashion those days and another popular classic at the time was "The Laughing Policeman" which also used to accompany the seaside slot machines that activated a laughing red-faced policeman puppet. We were easily amused in the 1950s. My grandmother also had an upright piano in her front room though I never heard anyone playing it.

Since there was no television in most homes, the most popular form of mass entertainment was the local cinema or "flea pit" as they were affectionately known. Boston had at least 3 main cinemas in my childhood and I was a regular visitor at two of them, the Odeon and the Regal. The other cinema, The New Theatre, made way for the Marks and Spencer store that now stands in its place. Unlike today, most cinemas at the time were more like a theatre with stalls and a gallery. You had to pay a bit more to go upstairs and there was also segregated seating in the stalls where the better seats had a premium price. The cheaper seats were known as the "Twopenny" stalls, the equivalent of less than 1p in the decimal currency introduced in 1971. My grandfather used to take his wife "Grandma Langford" to the pictures as a special treat very occasionally as they did not have a lot of money. My Grandad was always looking to beat the system and on one occasion bought tickets for the "Twopenny" stalls but took Grandma to the more expensive seats. This wheeze backfired when the usherette picked them out with her torch, checked the tickets and forced them to move to the cheaper seats. This humiliation was more than Grandma could bear and, after sitting in these inferior seats for a few minutes, kicked Grandad, walked out and made her way home to Fenside Road on her own. She was waiting for "Pop" with her rolling pin poised and the menacing words "Right you've had your fun, now I'm going to have mine!!"

My Grandmother always made a fuss of me and gave me special treatment. As I sat on her knee in my childhood she would tell me about her Irish ancestry from the days of the potato famine and her Christian names were witness to these origins because they began with Olive Emma and continued with the names of Irish Politicians of the era, such as Frost and Sleight. It was the equivalent of naming your daughter after the Liverpool football team. Once my mum and dad got a council house of their own in Wyberton, the only entertainment came from the "wireless" radio we had with its massive Ever Ready battery inside and its glowing valves. The names of the radio stations in those days evoke some memories but I recall it was the BBC Home Service that I seem to remember listening to most, especially on Sunday when the memories of "Two Way Family Favourites" and Billy Cotton Band Show with Alan Breeze and his "Wakey Wakey" always brings the smell of the traditional roast beef dinner to my senses. Two Way Family Favourites was a request programme for the British Forces Abroad from all their bases throughout Europe, a legacy of the Second World War. Workers Playtime was another popular programme of those days which saw many of the later household-name entertainers rise through the ranks of these programmes.

My Grandma Langford would also visit us every Tuesday with some pie or casserole for my mum to heat up. Grandma's speciality was Rabbit Broth. On these visits she would also bring me my comics, the Dandy and the Beano and, in later years when I had learnt to read, the Hotspur with its stories of heroic deeds from the fictional "Q Team" who became a powerful force in soccer by inventing a technology called "Attractapell", a magnetic device which the players would have in their boots that could both attract the ball and repell it, leading to world domination at the sport. My cousins Mick and Steve, who lived just down the road, grew up on the "Dan Dare" comics with the evil green headed Mekon.

As I grew up and was able to go out on my own without supervision at an age which would be criminal in today's society, I would go to the "Mickey Mouse Club" at the Odeon on a Saturday morning. Films at that time were largely in black and white and would always be accompanied by the "Pathe" newsreels of the time – very patriotic and jingoistic reports of the world news of the past month projected from the reels of film that spun through the ancient projectors. The Odeon also had a "Mighty Wurlitzer" organ that would rise up from the depths to entertain people in between films. These Saturday morning adventures used to feature adventure serial episodes with heroes such as Flash Gordon overcoming monstrous odds to save the world. My own particular favourite hero of those days was "Brick Bradford" who would be left in a hopeless and deadly situation at the end of each week's episode only to miraculously survive all manner of disasters to continue his adventure the following week. As I left the Odeon every Saturday lunchtime, I acted out Brick's heroics as I made my way home on the big green Lincolnshire Road Car Double Decker that would be parked at the main bus stand in the Boston Market Place.

My Uncle Bill, Aunty Iris and my cousins Mick and Steve, the Langford family living just down the road at Number 3 Yarborough Road, were the Jones's of our street that everyone else tried to live up to. Their Humber Snipe "Bomb" was the first and only car in Yarborough Road when I was growing up and it was therefore quite natural for them to be the first home to have a television. Uncle Bill was my mum's favourite brother. He was naturally entrepreneurial and a risk taker. I imagine that when he was in the Royal Navy he would have been their equivalent of Dads Army's Private Walker character or Del Boy in Only Fools and Horses. He was in many ways the opposite of my dad who had respect for authority and strong moral values but we all got on very well. My father was not a natural sportsman but he would join us kids and Uncle Bill to play football with us in the street or in Bowser's fields across the railway line.

It was my regular treat on a Tuesday tea time to visit them to watch Popeye on their newly acquired black and white TV with its tiny screen powered by a cathode Ray tube. Apart from the many episodes of Popeye, with Bluto, Olive Oyle and Sweet Pea that I watched, I have an abiding memory of seeing the football results one Saturday afternoon as the FA cup results were read out. "Derby County 1 Boston United 6 – I will repeat that Derby County 1 Boston United 6". This was a time when Derby were around the top of the old Third Division North whilst Boston were a successful non league team playing in the Midland League against such teams as their old rivals Peterborough United. This result still ranks as one of the giant killing feats of all time. Boston United's team had several ageing ex-Derby County players in the side and for them this was payback time.

Those early days of television did not broadcast 24 hours a day, as they do now, but only during peak hours and the rest of the time they displayed the Test Card on screen or occasionally, in between programmes, they would run a speeded up version of the "London to Brighton" rail journey to the accompaniment of the "Coronation Scot" theme music which I can still hear in my head as I type these words – "Da Da-da, Da-da-da-da-da....." Bonanza was a favourite in the Langford household and they used to call the wasteland that was their back garden overgrown with weeds and old creosoted railways sleepers "Ponderosa" after the Bonanza ranch. When it became time for us to be able to afford our own television, one of my favourite programmes was "The Lone Ranger" which always ended with a "Hi ho Silver" and "Who is that man – it's the Lone Ranger". This masked cowboy crusader with his faithful Indian Tonto would light up our lives every week.

Apart from the cinema, the other "out of home" entertainment that I used to adore was the funfair and the amusement arcades at the seaside. Every year Boston would be visited by the Annual May Fair and the

All the fun of the fair!

whole of the town centre would be filled with rides and attractions such as the "Spotted Lady" and stalls with coin operated slot machines, most of which were mechanical. I would regularly use the pennies my parents had given me to spend playing on the machines that flipped a ball into the air in the hope that it would fall into a hole that delivered a profit. Our summer holidays would involve trips to Hunstanton and Skegness and the arcade games on their piers that included a game which was powered by electricity and showed a montage of famous film stars, one of which you would press a button to choose before the machine randomly flickered through the choices. I always went for Ava Gardner because I thought she was the prettiest star on offer. My memory tells me that she rarely let me down and I often came away with a profit.

For my parents it was the weekly Whist drive that provided their entertainment at the local scout hut. I can't recall ever going to any music concerts with my parents but my first introduction to the stage came when I was in the boy scouts and joined the troupe of the annual Boston Gang Show which used to perform on the stage of Boston Grammar School. Before the age when my voice broke I was quite a good little singer and it was this singing prowess that landed me the part of Miss Susannah in a comedy sketch called Mr Brown and Miss Susannah. I was dressed in what could only be described as a very feminine girl's Easter outfit complete with bonnet and the song traced my transition from a very coy play-hard-to-get young lady being wooed by Mr Brown to a totally different bossy wife once I had got my man up the aisle. In a gesture that would be badly misinterpreted today, I was given a bunch of flowers and box of chocolates by the Manager of a local shop who thought I was a Girl Guide.

I used to love listening to the Goon Show on the radio with Michael Bentine, Peter Sellars, Spike Milligan and Harry Secombe. The BBC must have been one of the very first organisations to use technology to create the weird special sound effects these madcaps wrote into their zany comedy show. The BBC Radiophonics workshop that later went on to create the iconic Doctor Who electronic theme music must have required all of their creative genius to dream up sounds like "a gas cooker being driven backwards across the sahara desert." I loved all of the Goon characters and I used the very first piece of entertainment technology I

1970s entertainment technology with reel to reel tape recorder at Gadget Show LIve in 2011

ever owned, a tinny-sounding cheap tape recorder, to record myself imitating voices of characters like Eccles and his "You deaded me you dirty rotten swine", Neddy Seagoon and the old couple Henry and Min. This tiny tape recorder was a reel to reel recorder like its higher quality professional models. It didn't work too well but it gave me hours of pleasure.

The BBC also used to broadcast programmes like "Watch with Mother" for children. These were semi educational and did exactly what the programme implies, a reflection on the changes in society that have resulted in so many mothers being wage earners instead of bringing up young children at home. I can't remember whether they were in any way connected with the "Watch with Mother" programmes but I can recall Michael Bentine's zany programme for children called "The Bumblies". These were cute balloon like creatures that used to introduce themselves in one of Bentine's many voices by saying "I'm Bumbly One" and so on. At the end of the programme these little puppets used to float up to the ceiling where they went to sleep. I also used to watch Harry Corbett's puppets Sooty and Sweep and used to mimic Harry Corbett when naughty Sooty would do something wrong and he would say in his Northern accent "No Sooty No" all to no avail and it usually signalled the end of the programme when he would sign off by saying "Bye bye everybody, bye bye".

After my early love of my Grandma's gramophone, it wasn't until I had started at Boston Grammar School in 1959 that I heard any of the early pop records. There were lots of village fetes and sports days in those days with none of the powerful audio visual technology available today. The man who seemed to have cornered this particular market in and around Boston was Ron Diggins and his "Diggola". Ron used to run his business from home on the back road from Wyberton to Boston and his mobile PA and music system was operated from a small van with huge flared speakers mounted on top very much like the brass horns on the gramophone. He also had a bigger unit which he towed like a caravan for larger events.

As I got to the age of 14 and started to take an interest in both girls and music, my parents first bought me a "transistor" radio which I used to listen to in bed under the covers. It was in the days of Radio Caroline and Radio Luxemburg. The pirate station Radio Caroline was the basis of the recent film called "The Boat that Rocks" starring amongst others, Bill Nighy, but the real stars were the nostalgic records that featured in the soundtrack. One of the first records I remember hearing on the radio was "Johnny Remember me" by John Leyton. It stood out for me because of the echo effects on the backing vocals and these kinds of unusual and haunting record productions have played a big part in my life, especially Phil Spector's Wall of Sound. The first record player I ever owned was a red "Dansette" model and when I went out to get my very first "45" record of one of the hit records of the day I couldn't afford the six shillings and nine pence (about 30p today) that the original artist recording cost so I would go to Woolworths where they sold cover versions for considerably less.

The first actual original artist full price record I acquired was "My Guy" by Mary Wells, a record I still own today. I began to go to the dances that were held every weekend at Wyberton Village Hall where local groups would bash out rock and blues classics like "Loueee Louiii oh baby I gotta go". Like many other adolescents of my age, my shyness at approaching girls meant that I would wait until the slow numbers at the end of the evening to ask someone to dance and even then my choice wasn't based on who I fancied most, it was based on who I thought might be willing to dance with me and let me take them home. These were very innocent and happy times. As I grew older I became a big fan of groups like the Yardbirds and the Animals but it was the Phil Spector sound that got me completely hooked the moment I heard "You've lost the loving feeling" by the Righteous Brothers. I bought all of their albums as soon as they came out along with the other artists in the Phil Spector stable such as the Crystals and Ronettes and, later on, Ike and Tina Turner.

After I had started dating my first serious girl friend, Jenny Spinks, we started to go to the Boston Gliderdrome where there was a galaxy of the best stars with hit records. I have seen far too many of these acts to name but amongst the most memorable seen at the Gliderdrome are T Rex, Otis Redding, P J Proby, Ike and Tina Turner (who autographed her album), Righteous Brothers, the Move, the Equals, Cream, Geno Washington and his Ram Jam Band and the Crazy World of Arthur Brown playing their number one hit at the time "Fire". In later years there was a television programme called the Family Rock Tree which featured interviews with the famous stars of those wild rock days. Each of these rock legends was in their respectable middle ages by this time and I remember the episode which covered Arthur Brown and the bands which spun out of his Crazy World, such as Atomic Rooster. Arthur Brown still had extraordinarily long black hair and a beard which made him look like Rasputin. I finished up rolling on the floor in helpless laughter as he recounted in a very matter of fact manner the experiments he carried out to achieve his "flaming head" look as he was lowered onto the stage. He wanted to create the illusion that his head was actually on fire as he launched into his big hit and, his first iteration involved his mother's

Tina Turner at the Boston Gliderdrome

kitchen colander with candles stuck into the holes and attached to his head by an improvised leather strap. This proved unsatisfactory when the wax ran through the holes and burnt his scalp and the strap, when tightened, prevented him from opening his mouth to sing. His eventual solution was a custom made metal helmet which his roadie would spray with lighter fuel and ignite just before he went on stage. This also proved a problem on occasions when a combination of drink and drugs led the roadie to get lighter fuel on Arthur's clothes and ignite them instead!!

Other acts featured in this amazing series were the Nice and Yes. I have seen Nice a number of times in my life and have some of their LPs. Their talented organist Keith Emerson used to stick a dagger in the keyboard during numbers like their version of the West Side Story classic "America" and in this programme the bass player recounted how Keith threw the dagger into the speaker column and it bounced off to injure a member of the audience. Rick Wakeman told how he came to leave the successful band Strawbs to join Yes as their keyboard player only to witness their lead singer and bass player fighting on the floor at his very first rehearsal.

As I graduated from the cheap cover version singles available at Woolworths, most of my more upmarket purchases started to come from sections in Boston's two main record stores, Boots the Chemist and WH Smith. I was browsing through the collection in Boots one day trying to find a way of getting rid of the small amount of cash in my pocket when

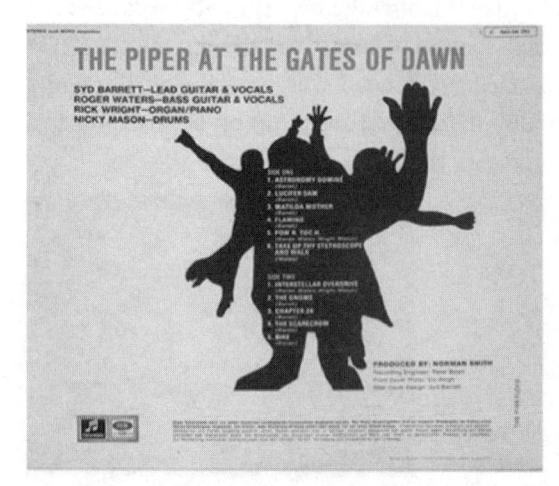

Pink Floyd, Piper at the Gates of Dawn

I came across the album that would bring me my favourite band of all time. I had never heard of Pink Floyd but the back cover of their album "Piper at the Gates of Dawn" and the names of the tracks like "Astronomy Domine" and "Interstellar Overdrive" was enough to persuade me to buy it and rush home to stick it on my little red Dansette record player. The sounds that drifted and blasted into my bedroom blew my mind and I went on to see them live no fewer than seven times over the years.

Pink Floyd were the real innovators of entertainment technology. They had a fantastic light show and sought to make their gigs truly memorable through not just their music but audio visual innovations. Their swirling light projections, using light projected through liquid gel rather like the lava lamps that became popular at the time, were a perfect backdrop to Sid Barratt's eccentric guitar solos. Their organist Rick Wright used a control on his keyboard they called "The Azimuth Co-ordinator". It was like a gearstick he could use to shift the sound around the speakers they used in the show to give the effect of movement. Three dimensional sound enabled them to give the illusion of movement and I remember watching them live at the Crystal Palace Bowl where they had speaker stacks not only on stage but right at the back of the bowl and they began the act with the sound of a helicopter flying over and around the audience whilst all we 1970s flower children audience craned our necks to see this invisible machine apparently above our heads.

My own early attempt to use technology to entertain was rather pathetic and a bit of a reflection on my limited engineering skills. I dreamed up the idea of running a fortune telling tent at a fund raising day one summer at our scout hut. I wanted to create this crystal ball with mysterious coloured lights that glowed and changed colour at my command so I devised a machine with a wooden cylinder to which were glued metal strips which acted as electrical circuit contacts so that when I rotated the cylinder with a little handle it would light up different coloured bulbs as the circuit connections were made. This device was cunningly mounted under the bottom half of a chemistry set flask that I had surgically modified with one of my dad's woodworking tools. The whole contraption was concealed by my mum's dining room table cloth and for good measure we made the inside of the tent as dark and mysterious as possible. Despite my best efforts and a heavy costume for disguise, people still recognised me!

Televisions in these early days had very small screens which were lit by a cathode ray tube and the electronics inside contained glowing valves and bulky circuits. The set would fairly frequently break down which would lead to "The Television Man" being called out to do a home repair. Mr Mellor was the television man in question and I would be fascinated as he diagnosed the problems and got the set back on his feet again. Like so many of the jobs that have now disappeared for good as technology has been consumerised as we moved into a "replace not repair" culture, it was a skilled task that Mr Mellor had prepared for over many years through a combination of apprenticeship training and experience.

My out of home entertainment, apart from Saturday nights at the Boston Gliderdrome dance hall, was mainly a regular Sunday night visit to the Regal cinema. The Regal was on West Street not far from Boston railway station and my first school, Staniland Primary school. Like all the decent cinemas in those days it had an upstairs and this is where I would head for the back row with my girl friend Jenny. By this time the old black and white films of my Brick Bradford heroic childhood had been replaced by colour films and Hammer Horror films were at the height of their powers. These horror films with Peter Cushing and Christopher Lee were fantastic entertainment not just for their fairly tame by today's standards scary scenes but for me, as a young adolescent boy with hormones starting to kick in, it was the eroticism of these sexy vampires that were a major attraction. My favourite horror film star was

The Regal Cinema Boston under demolition

Vincent Price with his wonderful rich voice and sense of enjoyment at being a villain.

There were many memorable evenings spent at the Regal and my all time favourite films of that era were Doctor Terrors House of Horrors, Dracula and Plague of the Zombies. I also remember seeing what might loosely be described as my first sex education film called "Helga". This was a German film dubbed into English and had only really got past the British Film Board censors with an X certificate because it had been produced to look like a genuine educational film with each sex position or foreplay technique introduced by a narrator dressed like a doctor with a white laboratory style jacket. I can picture the narrator now with his thick bottle lens glasses and brow glistening with sweat showing every sign of being turned on by what he was presenting in animated tones whilst the sub-titled translation and English language dubbing was very refined and unemotional.

Each of these evenings had a "B" film that was shown before the main film. These B films were low budget affairs, often shot in black and white, which followed a very similar pattern of the earth being threatened by some alien force and rescued by the skill and dedication of the scientist hero with whom some attractive woman falls in love in the closing sequences. There were classics with title like "The Beast from the Black Lagoon" and "Killer Ants from Space". I am convinced that the casts used in many of these were the same because the scientist hero always seemed to be played by Dana Andrews. The Regal, in its declining years as television took hold in people's homes, began to be used for other forms of entertainment including hosting the pop roadshows that used to tour the country in those days featuring about a dozen of the big stars who had current hits in the charts. The roadshow I remember seeing come to the Boston Regal had artists such as Gene Pitney (in the charts with Twenty Four Hours from Tulsa), Cilla Black, Wayne Fontana and the Mindbenders, Gerry and the Pacemakers and Them (with one of my favourite musicians of today, Van Morrison).

Like many teenagers of my age, I loved to watch live bands and imagine that I was there on stage with my heros. I especially enjoyed watching the bass players who, for me, were the unsung engine room of the band playing the riffs that the lead guitarists in the limelight could

John Entwhistle of The Who

exploit. I loved bass players like John Entwhistle of the Who, Jack Bruce of Cream, Roger Waters of Pink Floyd and Andy Fraser, the bass player from the band Free who's most memorable hit was "Allright Now". Each bass player had his own mostly undemonstrative style. John Entwhistle aka "The Ox" would be almost like a statue blasting out fantastic bass riffs like the one in "My Generation" whilst Pete Townsend in contrast would be a whirling destructive dervish ramming his guitar into the speaker stacks. Jack Bruce had a very jazz inspired style with really innovative bass lines whilst Roger Waters was a long haired menacing figure. Andy Fraser reminded me of a "Kelly Doll" by the way he used to rock rhythmically to the bass lines of "Alright Now".

Somehow I managed to find myself in a rock band at school whose driving force was a pupil one year above me called Tyrone Dalby. He played bass and I joined initially as a vocalist although I longed to learn to play bass guitar. I did have my own cheap acoustic guitar at home and was trying to learn chords using the tried and tested "Bert Weedon's Play in a Day" instruction book but the thin wire strings of my guitar played havoc with the soft skin on my fingers. I did however manage to persuade Tyrone to let me have a go with his bass guitar with its thicker strings and eventually I did get to perform on this in public on the one and only occasion at the Wyberton Scout hut when I played bass and sang on the Otis Redding hit "Midnight Hour". It was appallingly bad, not least because I couldn't master the art of playing and singing at the same time. Despite the publicity shot we got in the Lincolnshire Standard weekly newspaper we only did a handful of gigs, including one in the assembly

Me (at the top) as the lead singer in the under-rated band Gotham City

hall of Boston Grammar School when I borrowed my dad's old RAF uniform jacket and the "Chindit" hat he brought back from India after the war. I probably looked ridiculous but it almost certainly distracted the audience from the quality of my voice.

As I started my working life in 1967 with Post Office Telecommunications, televisions were still almost entirely still black and white. It was in 1969 when I was on a Post Office training course in Stone, Staffordshire that I witnessed man's first landing on the moon on a black and white television set in the "Mess Room". This Staffordshire location had been a military camp in former years and so the accommodation was based in wartime Nissen huts which were a mixture of semicircular constructions with corrugated roofs and prefabricated utilitarian rectangular blocks. As far as I recall, there were no individual rooms and we all had to share a dormitory and the bathrooms. It must have been a Spartan life for the soldiers who had been based there in former years.

In-home entertainment systems for playing music and radio were still very much a hobby pastime for the hi-fi "petrol heads" whose special language was full of "woofers" and "tweeters" and "Dolby". My interest in rock music led me to do my own research on what systems I might be able to afford to equip my room at the "High Hall" university hall of residence. I coveted the idea of top of the range Wharfedale speakers and

University of Birmingham High Hall

a Goodmans turntable with its counterbalanced stylus arm. I seem to remember that the amplifier I chose was also made by Goodmans. 1968 was the year that I began my three year chapter of life studying Electronic and Electrical Engineering. The university had a policy of trying to get all first year students into a hall of residence as for most students this would be their very first experience of life away from home. This policy also extended to trying to ensure that first years would have a double room to share with another student. I found myself sharing with a Leicester lad called Tom Wigfull whose father ran a business near Enderby just off the M1. I had never met Tom before but he was on the same course as me and although we were totally different personalities, we rubbed along with each other quite well and he got to know my musical tastes quite quickly. In those early days I used to catch the train home at weekends to play football so Tom had the room to himself from Friday to Sunday evening.

As the decade of the "swinging sixties" drew to a close, the technology used in recording studios continued to innovate to produce more interesting sounds than the exciting raw rock that had seen the "Mersey Sound" phenomenon. "Phasing" became quite a popular effect achieved by mixing the same sound through two channels and slowing and speeding one of the channels to get a futuristic effect. Although the Pink Floyd with its genius but errant lead guitarist and writer Sid Barratt were the real pioneers, it was the Beatles album "Sergeant Pepper" that hit new standards in studio recording production. I had been fortunate enough at the Freshers Ball in Birmingham in 1968 to see Pink Floyd Live by which time David Gilmour had come onto the scene as first a backup and then eventually a replacement for Sid as he went into meltdown. The supporting act at this ball was PP Arnold who had a hit with "First Cut is the Deepest" at the time and her backing band was an unknown (at the time) group called the Nice.

My musical tastes and love of the Pink Floyd brought some friendly

Pink Floyd Album
Ummagumma in 1969

banter between myself and a fellow student who became a good long term friend, Derek (Des) Wildman, who hailed from Sheffield but spoke with quite a posh accent. Des was a good classical guitar player himself and his idol was Eric Clapton whom I had seen years before when he was with the Yardbirds. He thought David Gilmour of Pink Floyd wasn't fit to be in the same room as Eric "Slow hand" Clapton. Des and I used to listen to music together in the High Hall room, sharing our first hearing of new albums we had each bought. He was engaged to a girl back home in Sheffield who was the love of his life and whom he would go home to every weekend. One Sunday he came up to my room in a state of heartbreak because his girlfriend had ditched him. I wanted to cheer him up and suggested that we should listen to some music together to take his mind off it so I put on the latest single I had bought that weekend. I could hardly have chosen a worst song in the circumstances as I put on the new Sharon Tandy single "Stay with me Baby" which quickly became accompanied by the sound of Des's sobs.

I continued to build my record collection and go to live gigs whenever Pink Floyd were in the vicinity. I watched them at Mothers Club in Birmingham at the gig where they recorded the live part of their double Album "Ummagumma" that included live versions of "Saucerful of Secrets", "Set the Controls for the Heart of the Sun" and "Careful with the Axe Eugene". The Floyd continued the exploration of sound technologies and worked with a producer called Ron Geesin to create a weird track called "Grooving in a cave with some Picts". Despite the fact that Sid Barratt had left to create his own appropriately named "The Madcap Laughs" album, there still seemed to be a legacy of his child-like lunacy in the band. I also saw the Floyd at Birmingham Town Hall where the highlight for me was their organist Rick Wright taking over the magnificent pipe organ at the Town Hall to create their trademark ethereal sounds.

My love of music, entrepreneurial spirit and desire to meet girls were the driving forces behind my decision to start my own mobile disco in my final year at Birmingham. Very unusually, the University authorities had allowed me to stay my whole three years in High Hall, the last two of which I was able to spend in a room of my own. My final year room was on the 17th and top floor of High Hall overlooking a mostly green suburban landscape towards Canon Hill Park. I was privileged to have a balcony outside my window which allowed me to climb out of the window and take in the inspiring sights below me. This trick was usually reserved for female guests that I wanted to impress. In the summer holiday before starting my third year I persuaded my dad to use his carpentry skills to make me a special chest high cabinet with two turntables mounted on top, a mixer/amplifier shelf and sections in its body to store my records and speakers. It was mounted on castors so that it could easily be wheeled around on flat ground. With this "babe magnet" in my armoury I sought to enjoy my last year at university "to the max".

My mobile disco was kept in my room and because it was always permanently wired, it acted as my in-room hi-fi system and every Sunday night I would wheel my kit out of my room and into the lift for my weekly discos in room LG4 on the lower ground floor. These discos featured mostly what are still today dance classics such as Tamla Motown and the Rolling Stones "I can't get no satisfaction" and became a very popular haunt for students across the campus. It was this success that led me to set up "The Progressive Underground and Rock Society" for which I recruited the legendary radio disc jockey John Peel to act as President. I would pay regular visits into the record stores in Birmingham City Centre to check out the latest releases and in those days you could use little booths in the shop to listen to a track before you bought the record. It was on one of these visits that I chose to listen to "Silver Machine" by an unknown band called Hawkwind. Like Pink Floyd, they were real innovators with sound and not only did I immediately buy the single, I made enquiries about booking them for the student union. I could have hired them for £50 had there been a suitable vacant slot in the room booking calendar. "Silver Machine" became a number one hit not long after and I was left rueing a missed opportunity.

Every week I would also buy the New Musical Express (NME) newspaper to keep up to date with new bands and new releases and in each issue the back page was always filled with an advert from a new music record mail order company called Virgin Records. They offered cut price albums for sale and had a massive list shown on their full page advert. The special "Rock Soc" film night that I had organised to show rare footage of Jimi Hendrix and Pink Floyd generated almost £100 in profits which I determined to plough back into my new enterprise by buying a big set of new albums from the Virgin records catalogues on the back page of NME. I was certain that my bulk order of around twenty albums ought to attract a higher level of discount so I rang the Virgin Records number and did my best to haggle for a better price. The person I spoke to put me through to "The Boss" who was the only one who could authorise such special deals and it is quite likely that the person who gave me a small extra discount was the Richard Branson who would go onto to extend his brand values into the airline, banking and mobile telephone industries.

My time at university came to an end and it was time to pack my disco gear into the back of a minivan my dad had hired to bring me and all my belongings back from Birmingham to my home in Boston. By this time my new career path had been set and I was destined for Leicester and my job as an Executive Engineer in charge of Internal Planning EI7 group. My first residence in Leicester was a very short term stay at a B&B in Milligan Road not far from the Leics Cricket Ground. This was just a temporary measure until I could find a place of my own. I bought the Leicester Mercury every evening to search for rented flats and thought I had found what I was looking for when I arranged to visit a flat at the Gynsills near County Hall out on the A50 but I was beaten to it by someone the landlord probably thought was less of a risk than a new arrival to the city of Leicester. Eventually, a short time later, I just happened to bump into one of my fellow Electrical Engineering students from Birmingham as he was walking down Granby Street in Leicester City Centre. It just turned out that he needed to find a rented flat as well so we joined forces and found a ground floor flat in a Victorian terraced house just off Hinckley Road on Gimson Road.

We each had our own bedroom and there was a big lounge, a cellar and a kitchen and bathroom in an extension at the back. The property's added attraction was the 3 girls who shared the upstairs flat. My flat mate was working at AEI just up Hinckley Road and had a fiancée whom he eventually moved in with after a few months with me. Whilst I was here, I began playing part time professional football for Loughborough United in what turned out to be their final year's existence before the ground was sold to make way for a new sports centre. Our Manager at the time was Richie Barker, a once promising young soccer star before a leg break ended his career. Richie was a very nice guy who later went on to manage Sheffield Wednesday.

During that year at Gimson Road I became engaged to Angela McFadyen from Barnsley and during the summer we set up home together in a rented flat in a farmhouse overlooking the Welland Valley at Dingley near Market Harborough. This move was to be closer to Angela's first job as Head of Physical Education at Robert Smyth Grammar School in Market Harborough. This was a peaceful, almost idyllic spot at the end of a long driveway just off the A427. Although I still had my hi-fi system, this was the only entertainment technology in the flat. We had no television and never really missed its absence. From Dingley we moved to another rented flat on the ground floor of a property in the leafy Stoneygate area of Leicester, within walking distance of the city centre where I worked. We still had only one car at this stage, a situation that persisted throughout our 14 year marriage.

It was one year after this in 1973 that we moved into our first owned property in Great Glen just south of Leicester. I took Angela to look at this 3 bedroomed detached when it was still under construction and I can remember her tears when she walked into what was to become our living room. The room looked so tiny in the empty shell that was still being constructed but we went ahead and realised as soon as the room was furnished that it was quite big enough for our needs. We bought our first colour television set here and my old hi-fi system was confined to very occasional use.

It was this house and its location that led to my re-introduction to the stage after the long absence since my Boston Gang Show and Gotham City rock band days. Our new home at 38 Cromwell Road was on a horseshoe estate that was still under development with the houses on Naseby Way still empty shells when we moved in. Our new neighbour in the house that stood opposite us with our respective back gardens backing on to each other was a single girl called Marion Sanderson. She had some friends who were members of the Kibworth Drama Group and she organised a very innovative garden party for her house warming. She invited everyone for a meal in the sun provided they brought along not only a bottle but a spade or some other garden tool to dig her garden for her. Like all new houses at that time, the gardens were used as a repository for all manner of rubbish before they were covered with top soil and given a layer of respectable turf. As one of her new neighbours, Angela and I went along with fork and spade and had a lovely afternoon digging and eating and drinking.

Like nearly all village drama groups, Kibworth was blessed with a surfeit of ladies and not enough male would-be thespians. "We need more men" was the siren call that led me to go along to the Kibworth High School Community Centre that was their home. The group was led by a husband and wife couple, Andy and Janet Munro who lived in Kibworth where Andy worked as a teacher at the school. The first production I watched was a spoof musical called "Dracula Spectacular" directed by Andy and featuring a cast of local people that would soon become my new social circle. In the first production that I saw was Janet Munro, Heather Wilson, Tom Grey, Gordon Dainty and Barry Hodby, who played an unlikely hero whose name I seem to recall was Nick Nicodemus. Barry was a lovely chap playing a tongue in cheek character in a performance that went beyond the description "wooden" to the extent that I thought he was a good actor deliberately pretending to be wooden.

I went on to do many productions at Kibworth including "Oh what a Lovely War" directed by Andy Munro, and various pantomimes and comedies that were this group's stock in trade. Tom Grey was the "ladies man" in the group, a large middle aged man with a beaming smile and a twinkle in his eye. There were too many memories to recount but although we all put our hearts and souls into our performances, it was a fun activity that allowed us to enjoy ourselves and give me license to play some of the practical jokes I became renowned for and that would have got me sacked at the Little Theatre. In one pantomime production in which all of us had to double up parts, I had the pleasure of being the back end of the traditional panto horse in which the front end was a local lady called Ann Perry who worked as a receptionist in a local dentist. Ann had a quick change from her previous dance routine where she wore a silky short dress and towelling knickers over her own underwear. I was already waiting in the wings in the rear end horse costume when she quickly slipped out of her red dress and, in her bra and pants, put on the front end costume and, bent down, with my hands on her waist, we went back on stage to do the horse's comedy routine which included some dancing. In the darkness of the horse costume I could see our legs swinging about as a close to being synchronised as possible and, without warning, I pulled her red towelling over-knickers down round her ankles to much hilarity and a red face from Ann.

On another occasion I was starring in a comedy called "Bed Full of Foreigners" in which my hero character found himself having to escape from forest wolves which had ripped his clothes to shreds before he succeeded in making it back to what he thought was his hotel room. Our simple set had a bed on stage and I had to strip to my briefs and climb into bed into which I would shortly be joined by the German Frau whose room it really was. This flimsy plot supposed that the German Frau was unhappy with her husband and was prepared to climb into bed with a total stranger. On the last night of the production, Tom Grey brought along a rubber object which looked like an apple until it was squeezed, at which point it turned into an erect penis. During the interval Tom sneaked this under the covers of the bed awaiting my scene. When I climbed into the bed I put this joke shop purchase between my thighs and awaited the entrance of the German Frau played by Lynette Rowbotham. As usual, she stripped to her bra and pants before pulling back the covers to discover me with the now squeezed toy looking like an erect penis between my legs. She pulled the covers back up in disbelief before sliding gingerly into bed beside me, now trembling with fear and unable to remember any of her lines.

On another occasion I was in a play called "Not Now Darling" with Heather Wilson in which I played the Manager of a fur salon and she played my cheating wife. This was another of the Whitehall farce type productions we all favoured and towards the end of the play, when I discover her wearing a fur coat and, suspecting her philandering, I order her to take the coat off. The plotline had her supposedly naked under the coat after being discovered in a tryst with her lover. With her back to the audience Heather protested that she couldn't take the coat off and when my persistence continues she opens up the coat as a prompt for my classic line "Good heavens my wife's stark naked". The real life Heather was a typical pale complexioned genuine redhead and on this one occasion I changed my line to "Good heavens my wife's not a natural redhead". I was a very naughty man in those days.

Many of the Kibworth productions were musical and although I never had a good voice once puberty had broken it for me, I never let a lack of ability stand in my way, especially in Music Hall productions and pantomimes where the songs are easy to sing. There was one notable exception however which came about when I was doing my DMS course at Leicester Polytechnic. Part of this course involved a 2 day residential session on "T group" sensitivity analysis in which you go through a set of exercises to help you see yourself as others see you. For many people this kind of training was very traumatic and we did have more than one person walk out during the two days. A lot of the time was spent in group sessions in which everyone would sit in a circle with our tutor sat in the middle to make sure things didn't get out of hand. The only rule was that you had to talk about the "here and now". These sessions always started in silence until someone "cracked" under the tension and made a comment of some sort that went on to trigger discussion.

In one of the sessions however we were split into pairs and sent off into the grounds of the conference centre where we had about half an hour to each reveal to our partner the thing we most disliked about ourselves. What the tutor didn't tell you was that when the group reassembled, each person had to reveal the same thing to the whole group. My palms were sweating profusely in nervous energy as it came to my turn and I revealed that I didn't put myself at risk by expressing my opinions to people and sometimes used that to my advantage in ways that made me look good at someone else's expense. The group's discussion about my weakness led to the suggestion that I should try to do things which put myself at risk of failure. My first opportunity to test this out was to audition for the male lead role in "The Pyjama Game", a musical which had such hits as "Hernando's Hideaway" and "Hey there, you with the stars in your eyes". My character in this romantic musical about a pyjama factory was called Sid Sorokin and I was required to sing songs which did indeed put me at considerable risk of failure and humiliation. The week of the show was a real test of character for me and an endurance test for the audience. I had to smile when I read the critic's review in the following week's Harborough Mail. She very generously described my singing in the show as "Eccentric" – a word which she must have pondered over for some time!

    I did get involved in the practical use of entertainment technology when I agreed to direct a spoof horror called "Curse of the Werewolf" written by Ken Hill. I was prompted to choose this play as my directorial debut because I had seen one of his other plays "The Hunchback of Notre Dame" performed some time before at the Phoenix Theatre in Leicester. I enjoyed his anarchic comic style so much that I sought out this play for one of our productions. Even for an experienced theatre director this play was a major challenge with many quick and varied scene changes from the opening graveyard scene to a castle prison, frozen lake and many other locations which had to be simulated through clever use of the stage and lighting. For this production I took responsibility for everything including sound and lighting and programme design. I put all my creative energies into making this production special. It became the world's first "Smellaround" (Surround smell) production as I hand-coated the insides of all the programmes I had designed with boiled garlic as a protection for the audience against werewolves. The owner of the Kibworth hardware shop where these programmes were sold as tickets for the show was not too pleased to have his premises reeking of garlic for several weeks.

    For the "Curse of the Werewolf" I managed to enlist two of my social friends, Dave Timson and Malc Lester into the production. Neither of them had any real interest in acting but Dave and his wife Sue, along with the Rentacrowd gang and Malc and his wife Diann had all supported me in previous shows and I suspect that they fancied a spell in the limelight. I can't remember what part Dave Timson played but Malc played a zombie who met a terrible fate when he was decapitated at one stage in the production. As well as the "Smellaround" innovation this show also included the frequent use of background music to create atmosphere in the same way as I had enjoyed in all those Hammer Horror movies some years before.

The show began with a scene in which the villagers carry a coffin through the audience to the stage graveyard where it is to be buried. I arranged the lighting to create an almost total blackout with just an impression of the moon projected onto the back wall to give an eerie light. The school hall where we performed had no permanent stage as it was used by the school as a gym and workshop space in term time so every production required the stage to be built from big wooden blocks during the weekend before any show. I had organised the stage layout and lighting to create different levels including a flat area at audience level. The only available space for the sound and lighting rig was on a specially erected scaffold above an exit door where I was positioned like a master conductor ready to entertain the hundred or so people who could be crammed into the remaining space.

As the show started I played the beginning of a Pink Floyd track from the album "Dark Side of the Moon" which comprised a heartbeat growing louder and acting as a cue for the gravediggers to bring in the coffin and make their way to the stage. For good measure I had written an introductory monologue which I spoke in a German accent to add a sense of gravitas to the legend of the werewolf. On the evening that Rentacrowd made their presence known, Dave Timson was acting in a skating scene on the pretend frozen lake in front of the audience at floor level. This ice skating was simulated by the actors wearing socks on a polished floor and exaggerating the movement of old fashioned skating with hands clasped behind the back. On conclusion of this scene my Rentacrowd friends led by my squash playing solicitor friend Julian Branston held up their programmes and shouted out scores like 6.5 and 6.0 as marks for the skating. On the last night of the show I took some paper streamers up into my lighting rig and, in the darkness of the opening scene, threw the unravelled streams of coloured paper into the air so that they drifted softly down on the audience's heads to much screaming and mayhem.

There were many happy days and many friends made at Kibworth including Bud and June Edmondson who lived in a little bungalow just on the main A6 as it passes through Kibworth. We had many parties at their house. Bud was the Kibworth Golf Club local professional and a very good golfer who was a regular partner of Sean Connery in Pro-Am golf tournaments. He told us many anecdotes about his time playing at the highest level in the British Open and about the golf scene in the James Bond film Goldfinger which was shot at one of his former clubs. Sadly Bud had to quit his job through a back injury and eventually set up his own heritage club making business in Market Harborough. I am still the proud owner of two reproduction vintage clubs that Bud made for me.

At home in Great Glen, our audio visual equipment was enhanced by the latest Amstrad 100 midi hi-fi system which was introduced in 1976. This had a turntable, radio and two tape decks which allowed you to copy from one tape cassette to another and from the record player to tape. The medium of CD (compact disk) had still to make an impact on the market. I used this system to make up my own compilations for parties and also create "pop quizzes" in which our guests used to have to guess the artist and song name from the few bars of introduction. The hardest ones for most people were the catchy one-hit wonders that

Amstrad Studio 100 Midi hi-fi system

everybody could hum but rarely remember the artist's name. Favourites included "Wooly Bully" by Sam the Sham and the Pharoes and "It's good news week" by Hedgehoppers Anonymous, a band formed by ex-RAF guys.

It wasn't until 1985 that tapes and vinyl recordings began to be replaced by the new medium of Compact Disk. The first artist to sell more than a million copies of a recording on compact disk was Dire Straits and today even this medium is in decline as digital recordings are increasingly distributed in MP3 format onto flash disks and downloaded onto Apple iPods. As colour television and my squash and drama hobbies began to fill all my spare time, we rarely went to the cinema or to any concerts, apart from the occasional production at the Leicester Haymarket. The cinemas were in transition from the old two storey formats that modelled themselves on Victorian theatres to the multi-screen venues that are so popular today. One film I do remember watching in those days was the first Superman movie. As I watched Superman flying across the city skyline I had no idea that there had been a man pushing a trolley with Christopher Reeves lying flat in his Superman pose in the making of this special effect sequence and that this man would become a friend in years to come. The only form of out-of-home entertainment technology was a Space Invaders arcade machine that was installed at Squash Leicester. I have to confess that I became slightly addicted to playing this early arcade video game on almost every occasion I visited the squash club. I had no idea in those days that it would be video game technology that would become at the heart of my future career over thirty years later.

Space Invader Arcade Machine

They say all good things come to an end and the time had come for me to move on to a new drama group at the theatre in Market Harborough. Unlike Kibworth, the Market Harborough Drama Society had its own theatre, and I was cast in my first play "She Stoops to Conquer" as the "Young Marlow" and directed by the late Arthur Jones. I was playing the lead romantic male opposite Jan Wilson, one of the

Society's many good actresses, and alongside Ian Parry (whose wife "Queenie" had a professional acting pedigree) and Mark Bodicoat playing the scene-stealing buffoon character of Lumpkin. As this was my first part at the Society, I made a special effort to learn my lines and took my script with me to California when I was attending the Siggraph conference and exhibition. By the time I went to my first rehearsal, I was almost word perfect and stunned the rest of the cast by rehearsing without my book.

Like Kibworth, Market Harborough Drama Society holds many happy memories that could fill a book in their own right. Because MHDS had its own theatre with a 100 seat auditorium, permanent stage, lighting and sound, we were able to rehearse on stage almost from the very first rehearsal, a very rare luxury. The technology at the theatre was also very good with its own lighting and sound box at the rear of the auditorium and a good team of technical people which included Jeremy Thompson and his partner who I worked with later in a travelling production with an Arts Council funded play written by local man Graham Day and produced by Jeremy's Troubador Theatre company. This travelling show has the distinction, amongst the many plays I have acted in, as being the only occasion where there were more people on stage than in the audience. This happened in the quaint little town of Uppingham on a Saturday night when there were clearly more attractive entertainments on offer elsewhere. One of my fellow actors at MHDS had an unwelcome reception during his crucial opening speech. Tom Henderson, a very dapper and Scottish retired gentleman and a fine and much respected actor, walked up to the front of the stage to deliver his opening monologue when one of the two old and rather deaf ladies on the front row of the audience was heard to remark in a loud voice "Oh it's him – I don't like him!".

Ever since my first experience of live audience entertainment in cinemas, theatres and concerts, I have had a passion for the use of technology to augment and enrich performances. The audience dynamics of live entertainment are complex and fascinating but the most memorable productions are able to take the audience to "another place" where there is a suspension of disbelief. Each audience member is immersed in a world which is partly of their own creation, partly created by what is seen, heard and felt and also partly by the group psychology of the audience. As a general rule, in any kind of performance, including conference presentations, it is the first two minutes that are absolutely crucial. If you capture your audience and meet or exceed their expectations in this first 120 seconds then they will forgive almost anything that follows. If the crucial period goes badly, it becomes a real battle to win the audience back.

The technology used in professional productions has massively improved the quality of the London theatre experience over the last forty years, especially in stage musicals where the production is a stage version of a film. In 1972 I went to London for my first marriage's honeymoon and we saw two very contrasting productions. The first night was Jesus Christ Superstar in which Anthony Head played Judas and the second night was the controversial "Oh Calcutta" production which was largely a series of comedy sketches in which everyone was naked. Neither of these shows made any innovative use of technology although nudity did have its attractions for the audience. We went to the 1000th performance of Oh Calcutta and, as the cast took their bows, part of the scenery fell over and knocked two cast members into the audience. Because it was the 1000th performance, many people thought this was a deliberate part of the show until the stage manager came on and ushered everyone out.

Some fifteen years later, I managed to get tickets for the first week of a new musical called "Time" which starred the evergreen Cliff Richard. This was the first time that I experienced technology used to create special and physical effects in a theatre. The basic theme of this musical revolved around time travel and when the time machine "took off" on stage with much use of dry ice to create a smoke effect, the whole auditorium literally shook as a result of powerful speakers embedded throughout the auditorium. They also used a holographic effect on a huge suspended and disembodied polystyrene head that was lowered down to hover above the audience and onto which was projected a hologram of Sir Laurence Olivier's head delivering the Time Lord's dialogue. Unfortunately, probably because they were still in the teething trouble stage, the holographic projection was not precisely aligned with the polystyrene head making it appear that his nose projected from the side of his face. Special effects that go even marginally wrong can detract badly from what otherwise would have been a great show. Even the Pink Floyd's legendary use of sound, lighting and special effects sometimes did not go to plan and their intended effect at the Crystal Palace Bowl of an Octopus like monster rising through the mist out of the lake that was in front of the stage fell flat when the dry ice only produced thin wisps of smoke instead of a blanket of fog. The net result was that the two frogmen tasked with inflating this monster from the deep were in full view of the audience instead of being obscured by the mist.

Because MHDS had a very experienced team of backstage people on sound and lighting, I did not become involved in any work behind the scenes. Like many budding thespians, I enjoy being on stage so much that I am delighted to be looked after by people who love to do sound, lighting, props, costumes, front of house and so on. It is not the people who perform on stage that should get the applause and limelight, it is all those unsung heroes behind the scenes who make life so easy for the rest of us. The one exception to my lack of involvement in entertainment technology at MHDS was an Old Tyme Music Hall show that I co-produced with Ian Joule. This show was Ian's brain child and was intended to combine the history of Music Hall with the history of Market Harborough. Ian and his late ex-wife Jean came from theatre backgrounds. He started up what is now an international clothes brand called "Joules" and is based on "countryside" activities such as polo playing. His shop still stands on Market Harborough High Street alongside Joules cafe, the walls of which are covered with memorabilia from the days when Music Hall was at its peak. The business has been developed by his son Tom to be what it is today and much of their manufacturing is done in their own factory in China.

Ian had his little office in the back yard of Joule's cafe and shop and ran a small antiques business from a shed in the same yard. When Ian came to me with this idea of linking the two histories together in a theatrical production, I was very enthusiastic and more than willing to use Mass Mitec facilities at the White House in nearby Lubenham to create an audio visual component to act as a backdrop to the songs and sketches. I also did the research into Market Harborough's history and worked with Ian on the selection of suitable material that could easily be linked to key moments in the histories of both themes. Music Hall started in the London eating houses and pubs where the young men would perform to each other in the late 1700s. I linked this to some old prints I found of Market Harborough in that period when it was a major halt on the London to York stage coach run.

I wrote and narrated the historical timeline and also did a couple of sketches and monologues, one of which was linked to the history of the famous Foxton Locks and the inclined plane built in 1912 even as the canals were in decline as an industrial force. To accompany this piece of local history I performed a monologue from around the same period. The monologue was called "The day I saved the barge" and was a comic routine based on the notion that the horse towing this barge (as they did in those days), was an ex-Grand National winner which bolted and was dragging this barge at high speed towards its destruction when my character managed to do a tightrope walk along the tow rope and uncouple it from the horse. Ian and Jean had been separated many years when she died at a time when she was in the middle of rehearsing for what would have been her first play at the Little Theatre in Leicester.

I worked with many actors and Directors at MHDS in my time there. I was the stage husband of Jeanne Moore on several occasions including a George Bernard Shaw play called "What every woman knows" and the fabulous Arthur Miller play "Death of a Salesman" in which I had the privilege of playing the lead part of "Willie Loman", one

Me as Pop Larkin in a scene from Darling Buds of May with Ma Larkin played by Debbie Neath

of the classic male roles in the theatre today. Being an amateur actor has many advantages over professional theatre and one of the most important ones is that you can get to play leading roles that professional actors would die for. Willie Loman was one such part and I would go on to play other fabulous and varied roles such as Rene Artois in "Allo Allo", Pop Larkin in "Darling Buds of May" and Barney in "Last of the Red Hot Lovers" written by Neil Simon about a happily married man who has never known sex with anyone except his wife and after twenty seven years of marriage wants to have just one afternoon of lascivious pleasure with another woman. In the MHDS production of the Neil Simon play I played opposite Jeanne Moore who was a hardened serial man eater, Lisa Whitcombe whose role was a pot smoking hippie and Di Rickard playing "the most depressed woman in the western world".

The thing I love about playing these characters on stage is that I can legitimately explore aspects of my own personality that would be difficult or unacceptable in real life. Willie Loman and the other Arthur Miller character I went on to play in the Little Theatre production of "The Price" are very real human beings who find themselves trapped in situations beyond their control where failure is the inevitable outcome. Having worked as a salesman in IBM for five years I had seen older men whose best days were behind them trying to pretend to themselves, their colleagues and their family that they were successful and leading a fulfilled life. I had also seen the hurt experienced by my father when, after he had given his heart and soul to set up a new Day Centre for the Mentally Handicapped in Spalding, the authorities gave the Manager's job to an inexperienced young graduate. Having the chance to play Willie Loman allowed me to express my own hurt and anger at the torment such situations bring people.

I played a number of pantomime roles over the years of my time at MHDS, one of which was the evil Abanazar in the panto "Aladdin". My job was quite simple – to get the audience to hate me and "Boo" on every occasion I came onto stage. I achieved this on my first entrance when I burst through the closed curtains and yelled "Arghhh" at the top of my voice. On one night I noticed that when I did this a woman sat about three rows back in the middle visibly jumped into the air. I made a mental note of her position and later in the show when I entered the auditorium at the back and walked down the side aisle to the sound of audience jeers, I waited until I had got to the front middle and bellowed "Arghhh..." at the top of my voice in the general direction of where I remembered the woman sitting. At the sound of my voice she leapt out of her seat and shouted out "Oh shit!" much to her embarrassment and everyone else's amusement. During the same panto run, my second wife Carol took our grand-daughter Amie to see her first panto production at the age of three and when I shouted at the audience the first time in the show I heard a little voice start to wail in distress and knew immediately that it would be Amie. She was inconsolable after the show and kept saying "Grandad shouted at me" and crying. It took about 2 weeks before she would even talk to me again.

It was during my time at MHDS that London stage productions like Starlight Express and Phantom of the Opera were beginning to use innovative technology for staging and special effects. These were still very early days compared to what is commonplace in London shows today and will be regarded as tame when some of the immersive technologies featured in later chapters begin to find their way onto the stage. Market Harborough was and always will be restricted by the physical size of the theatre and its stage but it was whilst at MHDS that I got my chance to go on the big stage at the Leicester Haymarket alongside professional actors such as David Shaw-Parker in the pro-am production of "Follow the Man from Cooks" about the life and times of Thomas Cook, whose story was to play such an important and motivating part in my life.

I replied to an advert in the Leicester Mercury and went to the auditions in a rehearsal room not far from the city centre. Everyone had to read several different parts from the musical show which had been specially written to celebrate 150 years since the very first excursion. The Director was a young man called Daniel Buckby. I auditioned with well over 100 local actors and singers, including some from MHDS. One of my fellow MHDS thespians who auditioned was Siobhan Moore whom I had already acted with at MHDS and whose mother Pauline was also a splendid actress. Siobhan later went on to be a professional actress and played lead roles in the West End such as "Casper" in the musical version of the film "Casper the Ghost". As well as being a talented actress she had a fantastic voice. I played a number of different characters in the show including the Manager of the Leicester Work House – in institution that supposedly cared for the poor in exchange for their slave labour. In my small scene I had to sit in a mock "theatre box seat" at the side of the stage and eat a hearty meal whilst my poor inmates starved all around me. For three weeks, at the side of the stage just seconds before I was due to go on, I would be given a perfectly cooked hot meal of sausages, mashed potatoes, peas and gravy to eat on stage. The timing was perfect and I had enough time on stage to finish the meal before the scene finished.

Other memorable productions during my time at MHDS were "Blithe Spirit" the Noel Coward play which was directed by Jeanne Moore, "Candida" by George Bernard Shaw where I met Nadine Scott for the first time (she is now Nadine Beasley) and worked opposite Alison Parkes with whom I would also appear in "the Widowing of Mrs Holroyd" by D H Lawrence. It was in this play that my horrible character was killed in a mining accident and I had a whole twenty minute scene laid bare-chested and supposedly dead whilst they washed my corpse. In small theatres like Market Harborough, the audience is so close that appearing not be breathing for that length of time is almost impossible. I also played the beast in "Beauty and the Beast" and was very proud of my dance with Lisa Whitcombe's Beauty in a carefully choreographed ballet to the backing of the Disney film version of the "Beauty and the Beast" theme song.

During my time at MHDS, entertainment technologies both inside and outside of the home continued to advance. The "must-have" hi-fi system that everyone with some money seemed to aspire to was Bang and Olufsen (B&O), not only for its sound quality but its elegant styling. Midi systems were becoming very popular and this trend of technology increasingly embedding human expertise within itself, even in systems like my low budget Amstrad Studio 100, lessened the need to spend time researching and understanding the fundamentals of putting a quality system together. In many ways this also reduced the amount of enjoyment hi-fi buffs got from debating and researching the components that could form their own dream system to show off to their friends whilst talking knowledgeably about dynamic equalisation and Dolby surround sound.

In the cinema, SFX techniques using computer graphics were starting to emerge. The older SFX techniques that were used in films like Superman (you'll believe a man can fly) used "green-screen" photographics that enabled all the studio equipment to disappear to be replaced by whatever background the Director wanted and Christopher Reeves was being pushed around the studio on a trolley by a young SFX trainee, Bob Bridges, whom I would later work with on community film projects in Market Harborough.

It was one December in the late 1990s that I was approached by Celia Mackay, wife of the Richard Mackay that I had so nearly developed a joint venture with during my time in IBM. She asked me to meet her for a coffee and a chat at the Three Swans in Market Harborough and during this meeting used sufficient charm and flattery to persuade me to audition for a Ray Cooney farce called "Run for your Wife" at the Little Theatre in Leicester. The play was being directed by Roy Smith and starred John Moore as a bigamist taxi driver who has a hapless lodger played by the late John O'Malley, a man whose laugh was so infectious that many theatres would willingly pay for him to be in the audience. Celia and Lynette Watson played the 2 unsuspecting wives living in Wimbledon and Streatham respectively. In the flat above one of John Moore's residences lived a gay man with a preference for wearing pink shower caps whilst he was painting and decorating, and this was my first introduction to the legendary Ken Milton. I auditioned for the part of Detective Sergeant Porterhouse, a role that had been played by Eric Sykes in the original London production.

The play was due to open at the end of January and we only had 3 weeks in which to rehearse because John Moore, who was a very talented Musical Director and keyboard player, was already committed to the preceding pantomime over the Christmas period. This play was the start of a new chapter in my amateur dramatics life and another step up the ladder of professional production quality. The Little Theatre is one of the UKs leading amateur theatres with an auditorium that holds just under 350 people and has full time staff including a Theatre Manager, Set Designer and Wardrobe Master. Rehearsals were great fun with such a talented cast and there was one incident which will live in my memory forever. The plot of this farce involved the investigation of a traffic offence which John Moore's taxi had been involved in and this investigation drew in the two Detectives from the respective "manors" of Wimbledon and Streatham. My opposite number playing the other Detective was Adam Watkiss, a young man from Wigston who was also new to the Little Theatre. Adam came onto the scene in the second half of the play when we had a confrontation in which our dialogue consisted of a repetition of "Who are you then?" as we faced each other off stood almost nose to nose. On this particular night one of us started to smile at the absurdity of the situation and this started us both off helplessly giggling with shoulders heaving and tears of laughter streaming down our faces.

Adam was a wonderfully talented singer with a very charming personality. He later went on to appear in one of the very early reality television talent shows called "This is my moment" in which six unknown singers each performed one song on Saturday night peak time television and the audience used their phones to vote for who they liked best. This business model of telephone voting has now become an important part of broadcasting economics because it generates revenue that fairly accurately mirrors programme popularity. This particular show was hosted by one of the Spice Girls and every contestant who made it through the auditions and the week's coaching prior to the show received 50 pence for every telephone vote. Adam won that week's show with the largest vote of any of the contestants in the whole series and as a result of this he took home over £150,000 and entrance to a whole new level of show business that took him to the London West End, his own solo album, the Royal Variety Show and a UK tour.

After the success of "Run for your wife", I went to the Little Theatre general auditions which are held annually and give each of the theatre's play directors a chance to look at new talent coming into the Society. It is a good scheme and most play directors are only too keen to bring new faces to the stage. I am always very conscious that in small drama groups you see the same people in starring roles all the time and it gets boring for audiences. I followed up "Run for your Wife" with another policeman's role in Joe Orton's black comedy called "Loot" which is about a bank robbery in which the robbers break in from the basement of a funeral parlour and are forced to hide the money in a coffin. Loot was directed by John Ragg and I got the part of Detective Inspector Jim

Truscott, a died-in-the-wool Scotland Yard veteran who wasn't beyond using dubious tactics to nail villains.

Joe Orton, who was one of Britain's best known and most controversial playwrights, came from Leicester and actually started his career at the Little Theatre. He was openly homosexual and was eventually murdered by his male lover. John Ragg managed to get Joe Orton's sister to come to one of our rehearsals to talk about her brother and her memories of both him and the writing of Loot. My character had actually been specially written for Kenneth Williams who, although he became best known for his camp comedy roles in the "Carry On" films, was a fine actor. This play also included Douglas Main and Jacque Hamilton, a fine actress who has become a life-long friend. There is one scene in Loot where my character has to pull out a pair of false teeth as evidence and Joe Orton's sister told us that when the play was being premiered in the provinces, their father died and Joe came back to Leicester for the funeral. Joe seemed to have a morbid interest in his father's body as it lay in the coffin and took away with him a pair of his father's false teeth. That night, just as Jim Truscott was about to go on stage, Joe thrust the pair of false teeth into the actor's hand and said "See what you can do with these!" and that apparently is how the teeth came to be written into the play. She also told us that there were daily re-writes of sections of the play before it reached the West End.

My time at the Leicester Drama Society (LDS), the resident company based at the Little Theatre, has provided lots of opportunities to learn new skills and develop techniques which have often helped me in conference presentations. Not long after Roy Smith had cast me in "Run for Your Wife", he asked me to be Master of Ceremonies of an Old Tyme Music Hall with a difference. I not only had to introduce the acts, I was also involved in many of them. The show was written by Roy Smith and Ken Milton and had a magic act in which I played the great Chinese Magician "Won Hung Lo". The act started with me walking across the stage in my MC dinner jacket outfit behind a cloth held up by my two assistants, Lotus Blossom and Cherry Blossom, and emerging the other side of the cloth in a Chinese Mandarin outfit to begin my tricks.

Lotus Blossom and Cherry Blossom were in identical outfits with the usual basques and fishnet tights. Lotus Blossom was played by the

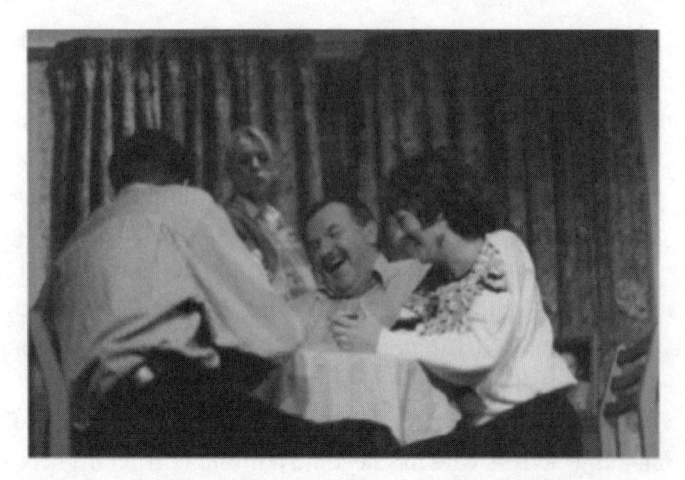

Having a laugh with Jacque Hamilton in the LDS production of the Alan Ayckbourne play Seasons Greetings

attractive, young and slim Elaine Toft whilst Cherry Blossom was played by the less attractive, not so slim pensioner Joan Daft. I had to learn a number of genuine magic tricks that were very familiar on television at the time, climaxing in the famous sword cabinet trick in which a cabinet was wheeled onto the stage and into which I climbed, still dressed in my mandarin outfit. My two oriental assistants would then brandish some sword sticks which they shoved through the holes in the sides of the cabinet till it would be impossible for anyone inside to survive. Lotus and Cherry Blossom spun the cabinet round to show there was no way anybody could escape but revealed a mandarin outfit clad magician clinging on to the back of the cabinet. The audience were still laughing at yet another trick that had either gone wrong or been exposed when I strolled on from the side of the stage in my original dinner jacket MC gear. It was during this show that I that I also had the rare privilege of singing with Adam Watkiss in one of the last productions he did at the Little Theatre before his big break on reality television.

Another skill that I acquired without mastering in my early days at the LDS, was ballroom dancing. I was in a comedy called "Last Tango in Whitby" in which I played a ballroom dancing teacher who falls in love with one of a party of pensioners on an annual holiday to Whitby.

This was a very funny play which included many scenes in which ballroom dancing played a large part in the action. Although, like singing, dancing is something which I love doing without necessarily being very good at it but the whole cast had a weekly ballroom dancing lesson on one of our rehearsal nights, being coached in some of the most popular dances featured in the play including the waltz, foxtrot, military two step and square tango which became my favourite dance of all. It was during these rehearsals that on the very rare occasions that I moved in near perfect harmony with my love interest in the play, Jenny Page, that I could understand why ballroom dancing is a passionate hobby for so many people and why the "Strictly Ballroom" programmes on television are so popular today.

It was whilst I was rehearsing Last Tango that I was commissioned to do some Harvard Graphics training in Scarborough, which is just down the North Yorkshire coast from Whitby. I had become fascinated by the many different actual locations mentioned during Last Tango that I wanted to see for myself if these places existed in real life and what they were like because, although Last Tango is a comedy, it is so well written that you get drawn into the lives of the characters and the places where the action took place. This included "The Royal Hotel" where the pensioners had the holiday booked and the Magpie Café by the river estuary. So, on the day my Harvard Graphics training was scheduled, I set out from home very early to have a look round and, in the early morning sun, I took my English Breakfast at the Magpie Café and drove up to see the Royal Hotel on top of the cliff. I fell in love with Whitby and the little village of Goathland which achieved television and film fame in years to come for its role as Adensfield in the TV series Hearbeat and as Hogwarts station in the Harry Potter films.

It was during this period that I had my first experience of 3D cinema and some of the special effects that were unusual in the 1990s but are now commonplace. I used to take my extended family to Florida to the attractions in Orlando at Disneyland, Epcot and MGM Studios. There was a 3D cinema experience in Epcot in which the audience needed to wear the red and green lensed cardboard glasses and would witness objects flying towards them. My particular favourite rides were Space Mountain, Back to the Future and the Tower of Terror. All of these rides were based on similar principles to that early experience of the Jules Verne Astroglide on Skegness Pier in 1981 but, as technology improved, the experiences became richer and richer and more and more available to the consumer market.

Yorkshire is one of my favourite counties and, possibly because of the number of years I was married to a woman from Barnsley, one of the regional accents that come most naturally to me. This accent was put to use in another Yorkshire based comedy, John Godber's "Up n Under", a heroic tale of a bunch of misfit rugby league players assembled after a drunken bet to challenge for a local seven a side cup. It was a lovely warm northern story with some great characters and was later made into a film starring the late Brian Glover whose Yorkshire accent was best known for his voice over in the Tetley Tea Advert that went "Tetleys make tea bags make tea". Brian Glover had also starred in one of the "Alien" films but actually started his acting career playing a Physical Education teacher in the film "Kes". He was genuinely a teacher in those days at a Barnsley school where one of my sisters-in-law went. In these days, as well as teaching, Brian used to moonlight as a professional wrestler, often fighting on the continent in France or Holland at the weekends. In Barnsley in the 1970s, France and Holland were metaphorically other planets beyond the knowledge or comprehension

Motivating my players in the LDS production of John Godber's Up n Under

of the Barnsley locals Brian used to drink with in the pub on a Sunday night. When these locals used to ask him why he wasn't in the pub on the Saturday night he used to tell the truth and say "Ah've bin wrestlin on t'continent" to which they would laugh and reply "Eeh tha's a raight one Brian". It was "Up n Under" that brought me onto the Tigers Rugby training ground in Leicester for a publicity shot in a scrum with the England Rugby Union team's "ABC" front row of Cockerill, Roundtree and Garforth. Many of the Leicester Tigers Rugby Union first team actually came to see the show.

It was at the LDS that I also acted on stage with a certain Richard Burton who went on to become a tragic figure in national press stories. This Richard Burton was a young man who had just joined the LDS and happened to share his name with the late legendary wild man of cinema who married Elizabeth Taylor. We were both in a play called "Blue Remembered Hills" written by Dennis Potter and directed at the LDS by Stephanie Liggins who was also a Casting Director at the Leicester Haymarket at the time. "Blue Remembered Hills" is a dramatic story based on the cruelty of school children and all the child characters are adults dressed as and playing the roles of these children growing up in the countryside. Richard Burton's character in the play was "Donald Duck", a lonely misfit of a child that was bullied by everyone and finished up being locked up in a burning barn at the traumatic finale to the play. This was Richard's first major role in front of a large audience and he loved being part of this experience so much that it became an unhealthy obsession that took over his life and, when the play was over, left him with a void he could not fill. A few weeks after the show had finished he murdered the landlady of the Leicester home where he was the lodger and finished up being committed in prison. It turned out that he had schizophrenic tendencies and his care in the community programme had failed and led him to commit this terrible crime.

Happily there are very few sad memories from any of my LDS activities. It was through meeting Lynette Watson in my very first play, "Run for Your Wife", that I became involved in Television Extra work. Lynette already had a great deal of experience of TV and film work which started when she was a child actress in the original "St Trinians" classic Ealing comedy. Lynette had an agent and an Equity card which was a pre-requisite for getting any kind of "Supporting Artist" work in those days. She told me about her appearances in programmes like "Peak Practice" and, as our friendship grew, she involved me in a little professional troupe she ran with other local performers like Paul Large and Linda Smart. The contract bookings we used to get to do were what were mostly nostalgia evenings, counted as contracts towards the minimum number necessary to apply for Equity membership. It was through Lynette that I became registered with her agents in Southend on Sea and Leeds and started to get bookings for TV Extra work.

One of the very first joint bookings we got was for a BBC Two comedy being filmed in a disused multi-storey car park in West London. We had to use Lynette's car as a prop for a car boot sale scene and this started me on a path on which I had quite a number of parts in various TV soap programmes such as Doctors, Bad Girls, Dream Team and "The Bill". The most common type of part I got as a background artist was playing a policeman and I was never sure how to react when my fellow actors used to tell me that I really did look like a copper in uniform. I had one episode in Doctors, which was normally filmed in the studio at the BBC Pebble Mill in Birmingham, where I and a couple of my friends from LDS, Steve Findlay and Jane Towers, had to drive out to the filming on location at Lickey Hills country park just outside Birmingham. All three of us had been kitted out in police uniforms and I was driving everyone in my car through Edgbaston when we were overtaken several times on a speed restricted dual carriageway by different drivers who, as soon as they spotted our uniform, slowed down dramatically after they had gone past.

Television Extra work is exactly as it is portrayed in the Ricky Gervais comedy series "Extras". Everyone I know who does TV Extra work is desperate for an opportunity to say something on screen. This is known in the trade as "doing lines" and is the holy grail of the television extra. For me it is not the extra money that doing lines brings (the normal daily rate for a TV Extra without lines is £70), it is about the experience of being on television. On one occasion I thought that I had finally made it when my agent rang me as I was just about to drive away from Birmingham Airport car park and asked me if I "could do lines". I didn't hesitate in my reply and he told me that I had a part as a Customs Officer in Dream Team and that I should report to the Three Mills Studios in South London at 7:30 am the following morning. I couldn't sleep with excitement and set off at 5:00 am the next day. It was when I reported to Wardrobe that I got my first inkling that something wasn't right because they said I didn't need a full uniform as I would only be in the background. My suspicions were confirmed when I sat in the "Green Room" waiting instructions and got into conversation with a fellow actor who told me he had auditioned for the part of a Customs Officer the previous day and had been cast in quite an important role for the storyline about a Russian footballer being taken back to Moscow against his will and being saved by the very Customs Officer I had been expecting to play!

Over the fifteen years that I have been involved at the LDS, entertainment technology has followed the same pattern as all the technologies covered in this book and has been impacted by the convergence of all these technologies to deliver richer and richer experiences both in-home and out-of-home. In my childhood, entertainment was an activity that you had to use very human attributes such as imagination to appreciate, and technology played a very minor part in delivering that entertainment. Whether it was books or listening to the radio, it was your imagination and your self-generated immersion in scenes and activities that you painted in your mind that brought the pleasure and the experiences that go towards shaping your life and your personality.

Out of home entertainments, like the cinema and the funfair, used technologies that were unavailable or unaffordable at home and provided richer and different experiences to open up new worlds and horizons. As home entertainment technologies became consumerised, more and more of these rich experiences have been brought inside the home with superb quality large screen displays, cinematic sound and now 3D, all acting as challenges to cinemas and forcing out-of-home entertainment to constantly offer something beyond what consumers now have in their homes.

To meet these challenges, those whose businesses depend on engaging large numbers of people in entertainment activities are now looking towards the technologies covered in the latter part of this book and which have played an increasingly important part in the later stages of my life and my career.

CHAPTER SEVEN
When Will We Ever Learn?

Education is one of the human activities which has been most impacted by the technological advances I have witnessed in my lifetime, so much so that there is some evidence to indicate that there has been an almost Darwinian evolution in the brains of young "Digital Natives" brought about by these technologies. Educational technology is an area that has become increasingly important not only in the later stages of my career but also in society as a whole as we struggle to understand the implications of the tools being used every day by our children.

Learning has always been a great joy and passion for me. I believe that all of my life's experiences on the journey described in this book have taught me something and have influenced who I am and what I believe in. The technologies described in this book, right from the transportation technologies of chapter one to the latest immersive technologies described in chapter nine, have all played their part in the shaping, development and progress of education throughout history because all these technologies involve communication between human beings and the world around them. From the earliest homo sapiens, human beings have processed the information that their senses have communicated to them and have learnt from the actions triggered by this processing. Since most of my higher education and early career has been in engineering, I see the learning process in many ways as being analogous to the feedback loop control system that I cobbled together for my final year project at

Not a candidate for best dressed pupil of the year at Boston

Birmingham University. Like the machines that we human beings have refined almost beyond recognition, we receive inputs, we process those inputs using our programming and memory, and we act to create outputs which we then observe as feedback to further enhance the sophistication of our brain's control system. This process could arguably be described as the true essence of learning and the way this has been impacted by developments in technology may be a root cause of some of the societal problems we are facing today.

Babies and small children love to explore the world around them and are guided through the magical and exciting unknown that they are born into by their parents, by their genetic programming and by the reactions they get back from their actions. They love to poke their fingers and hands into everywhere just to see what happens and what they can learn from it. Where there is no consequence to their actions, there is limited learning and limited shaping of behaviour. In the very physical world of the caveman, there was a strong connection between action and reaction in dealing with other humans and the world within which they lived. Today, paradoxically, we live in the most connected society in the history of mankind and yet, in many ways, it is simultaneously the most disconnected society in our history because our actions, empowered by technology, can have consequences to other human beings and to the world around us without any noticeable impact on ourselves. This is a very dangerous and unsustainable situation to be in from a purely engineering perspective, especially if our aim is to build a stable and manageable social system that can form the basis of an equitable and sustainable future.

Education and learning since the dawn of time has been a facilitation process in which the learner benefits from the experience and knowledge of a guide, coach, mentor or teacher who helps to bring understanding, develop skills, commit facts to memory or modify behaviour. In all these roles, the educator uses tools to help the learner to acquire these desired learning outcomes and the best educators are able to adjust their strategies, tactics and tools to the individual needs of the learner. Amongst the earliest learning technologies used by mankind are books which a self-directed learner can interpret using their existing knowledge and experience and which a teacher can use to guide the learner towards the desired learning outcome.

In the world that I entered in 1949, machines required the intelligence, skill and understanding of a human being in order to operate at maximum effectiveness in all human activities, including education. The learning technologies I have seen developed in my lifetime have gone from a process of making mass automated education accessible to the world towards a process in which the intelligence and understanding of the machines is used to deliver personalised learning based on an ever increasing understanding of the individual learner. The implications of this change are massive and will impact all areas of society, not just education. This chapter is the story of my journey through those changes and my view of where this might take us in the future.

My parents and both sets of grandparents spent a lot of time with me right from when I was born. Until my mum and dad got a council house in Wyberton, we lived, as many young families did during those early post-war years, with my mum's parents in Fenside Road just about a mile from Boston town centre. I was lucky that my grandad worked on the railways and gave me the opportunity to travel to different places with him and grandma at weekends. My dad's parents lived a short distance from Fenside Road in Granville Street, and one of my earliest memories was when we all stayed at their house at Xmas and I remember being laid in bed watching the snowflakes fall, glistening in the light of the gas lamps which still used to light the streets in those days. Grandad Wortley died of bronchitis when I was only 3 years old but I am told that

Site of my old primary school Staniland in Boston

I used to sit on his knee and he would call me "Old chap" just like the Joe Gargery character I would later play in Charles Dickens "Great Expectations" used to call Pip.

Apparently I was a very well behaved baby who learnt to spell at an early age and one of my party pieces was an ability to spell diarrhoea at the age of five, a skill, according to Microsoft Word's spell checker that I haven't lost some fifty seven years later. My parents also told me that I learnt to read at an early age and they used to impress visitors to our house in Wyberton by getting me to read the headlines from the Daily Mirror to them. This once caused a mixture of amusement and embarrassment when the headline read along the lines of "Vice ring smashed". I went to Staniland Primary School in Boston and neither here nor at home were there any forms of educational technology employed. One of my favourite lessons was a class given by Mrs Derbyshire in which she used to read adventure stories about "Biggles" and "The Famous Five". I did well at primary school and seemed to be always at or near the top of the class, much to the annoyance of the mother of one of my schoolmates, Clive Wicks. His mum was very ambitious for him, even at primary school, and he was always trying to get better marks than me in our frequent class tests. On one occasion, as the results of the latest test were handed back to us, he shouted out "I've got 92 marks I've beaten him", only to be deflated when he learnt that I had got 96 marks. It was whilst I was at Staniland Primary that I won the "Brooke Bond" handwriting competition which, if you saw my handwriting today, would be considered a minor miracle.

I passed my 11+ exam first time in 1959 and because of my birth date, I was one of the youngest in my class when I went to Boston Grammar School. This was in the days before comprehensive education and pupils were segregated both by ability and sex. "Clever" boys and girls went to Boston Grammar and Boston High School whilst 11+ failures went to the Secondary Modern Schools of Kitwood Boys and Kitwood Girls. During my first week at Boston Grammar I found myself in detention after school one afternoon for "running in the corridor". I was always in the top stream or "A" form at Grammar School and chose to study the science based subjects for Advanced Level exams rather than history, geography or languages. We learnt French and Latin at Boston

Boston Grammar School

Grammar School and, rather unusually, had Russian language lessons as an option.

Although I was reasonably capable at school and found myself in the top 10 of the class in most subjects, my passion was sport and I played in the School soccer team in winter and tennis team in summer. We also had a Cross Country club organised by Mr Vass and I was a great source of irritation to him because although I never went to any of his after school runs because of my soccer practice, I became the school cross country champion and held the school under 15 cross country record for many years. On the occasion that I set the record it was a crisp winter day and our course took us along the road to the river bank and quite some distance along the bank before turning and heading for home. One of the teachers was posted out near the half-way point to make sure no-one took any short cuts or stopped for a cigarette and when I ran back through the school gates a good minute or more ahead of my nearest rival, no-one would believe I had completed the course.

Our Grammar School was founded in the 1600s and had many traditions, including a Latin school song, Floreat Bostona, and a special half day holiday each year to celebrate Beast Mart which was a legacy of the days when a big annual cattle market was held in the town. There was also an annual local knowledge competition with a small prize in which you tried to answer about 100 questions about Boston and the surrounding district. These were mostly geographical or historical questions. I remember one of my classmates being admonished for his answer to "Name as many places which include the word Bain". His answer was "There Bain't be one". It was my modest success in this school quiz that led to my first appearance on television in Anglia TVs quiz programme for schools called "Top Town Quiz". Four of us were driven to the Anglia studios in Norwich where we competed against Lowestoft. I was fascinated by the TV studio because all the furniture seemed just very basic with a rough table for each team covered in blotting paper. In the days of black and white television, these things did not matter because everything looked fine on screen. Our team won the contest and I think our reward was a five shilling book token each (about 25p today).

Boston Grammar School had a good educational reputation and generally sent around 4 pupils to Cambridge University every year. As one of the top students, I was selected to take the special Cambridge Entrance Exam which was one of the qualifying filters. My parents would both have loved me to have gone to Cambridge or Oxford and every year we would all have a small bet on the Oxford and Cambridge boat race. No-one prepared me for the ordeal that was the Cambridge Entrance Exam. It was totally unlike any test I had ever seen in my life and so traumatic that I still remember some of the questions today such as "How many molecules of Caesar's dying breath do you inhale every day ?" or "How many tons of coal would it take to replace the sun for a week ?". My feeble efforts at answering these questions were not helped by a leaky fountain pen that deposited blots of ink all over the paper. This failure was also compounded by myself being one of the few people in our school to fail the "Use of English" exam.

The only educational technologies in use at our school were books and the blackboard. Education was very much of the "chalk and talk" variety supplemented with the usual laboratory classes for physics, chemistry and biology. The teachers whom I remember most clearly are not those who were especially able to get the best results from me but the ones who were somewhat eccentric like "Joe" Gledhill our Latin teacher who used to gain attention from pupils by either throwing a board rubber at them or lifting them up by their sideburns.

It wasn't until I left school to begin my career as a student apprentice with Post Office Telecommunications that I began to encounter any kind of educational technology other than books or the blackboard. In those days there were early overhead projectors (OHPs) being used occasionally with often hand-written transparencies but the first video technology to really capture my attention was the Optical Laser Disk. At the beginning of one of our courses in the Post Office we were shown a video that was played back from one of these disks which were the size of a long playing record. The video was some kind of motivational production about inspiring human achievements and it wasn't just the content that made an impression on me, it was the impact that a well-made atmospheric video had on my interest level and retention.

During my time at Birmingham University, the main media were still books and the blackboard, supplemented by laboratory sessions in which we worked in pairs to set up electrical circuits and experiment with the electrical and electronic devices of the day. My Lab partner was called Perrin, a local Birmingham lad with a passion for collecting coins. He sadly drowned during my time at the University. He was known for being a bit absent minded and I recall one incident in the lab when I was disconnecting a potentiometer from the circuit we had been experimenting with when he switched the electricity mains back on and nearly electrocuted me.

It wasn't until I started work as a Tutor at the Post Office Telecomms Management Training College in Cooden Beach near Bexhill that I began to explore the potential of electronic technologies for education for myself. Most of our work with our students still used books and blackboards but we did make use of video at the college and had our own mini recording studio and editing suite which were mainly used in sessions to teach interview techniques and body language. In these sessions we would video the students doing a mock interview and then play it back to the whole group for them to analyse and comment on. I made a special video of my own that I used to illustrate the importance of body language. I filmed myself in a couple of scenarios that simulated a counselling interview and a disciplinary interview in which I deliberately used misleading and inappropriate body language. For the counselling interview I tried to appear aggressive and uninterested with gestures like looking at my watch and fiddling with my collar, and for the disciplinary interview I was smiling and relaxed. I then used these silent videos to show the student group and get them to observe any body language signals to draw conclusions about what the interview was about. I found this technique very effective and interest generating.

The BBC have been pioneers in educational technology from very early days and the "Watch with Mother" programmes in the 1950s and 1960s were a classic example of how visual communications can be used by a parent / coach / guide / mentor / teacher to help a learner understand, acquire knowledge or shape their behaviour. It is the human intervention that is critical in this process because it is the relationship between the learner and tutor and the ability of the tutor to personalise the learning experience that will determine its effectiveness.

It was whilst I was at the Post Office Telecomms College in Cooden Beach that I had my first experience of a serious game with the "Action Maze" Exercise involving a pack of filing cards which took you through a simulated disciplinary case. It was a kind of role playing exercise in which you chose the decisions that you would make in a given case study scenario which then led to an outcome that would be used to illustrate the implications of your decisions and deliver the better understanding

Early learning technologies at the Gadget Show Live in 2011

and behavioural change that was the intended learning outcome. This type of role playing learning approach can be very effective when used in the most appropriate way and it is one of the educational techniques that lends itself well to the use of modern technology.

I attended many training courses in my time in IBM between 1979 and 1984 and these days were the beginning of some of the early uses of audio visual and communications technologies with the use of 35mm slides, video and even some of the very early data projectors that have now become a standard office equipment item. In schools, the early electronic calculators were starting appear with a great deal of debate about how their use would hinder the progress of education because children would become lazy and not learn the fundamentals of numeracy.

By the time I left IBM in 1984 to set up my Mass Mitec consultancy, the very first desktop computers were beginning to appear, such as the Commodore Pet, BBC Micro and the Apple computer. All these machines were still very much early adopter and hobbyist technologies but it was their embryonic nature that really lent itself to the learning process, especially for those enthusiasts who saw the potential of these technologies and went on to make their fortunes. Man has a great capacity to learn through exploration of the unknown and through collaboration and sharing of ideas. It is this exploration and collaboration which is driving a major trend away from the dissemination of existing knowledge towards the generation of new knowledge. It is using the advanced technologies that we have at our fingertips today that is creating many of our educational and societal challenges.

It was during the period after the emergence of the early desktop computers that technology began to be more seriously applied to education. In these early days the focus was almost inevitably around using the traditional characteristics of machines to make education faster, cheaper and more productive. The ability to be able to create the electronic equivalent of books, firstly with text only and later with graphics, store them and distribute them in electronic format was the beginning of our efforts to understand and use these new electronic tools. In my own case, I was making active use of programmes like Execuvision, 35mm Express and then lotus Freelance, Harvard Graphics and finally Microsoft Powerpoint to inform and educate my clients and

Prestel screen

prospects. In larger conference venues I would be using 35mm slides because of their superior quality, and in my smaller seminars I would use the three lens Thompson data projector that was based on cathode ray technology.

In those early days before 1990 when the telephone network was hampered by the limited capabilities of the old copper cable network and modems were slow and unreliable, Prestel, developed by British Telecom, was a viewdata system that heralded the potential of the telephone network to make information available anywhere in the world. This system sent very crude text and graphics at a speed of 1200 bits per second out to the computer and received instructions on page selection at 75 bits per second but it did work, and I used to use it in the early 1990s to get the kind of information that television users would access by using Teletext or Ceefax service.

It was in 1985 that the CD-ROM disk became available for desktop computers and in 1990 when the recordable CD was launched by Phillips. This massive increase in storage capacity of a highly portable and low cost medium suddenly made the distribution of learning content a much more viable option and the ability to be able to record CDs on the office or home computer also opened up the potential for user generated content to be distributed electronically. It was in the latter half of the 1990s when other technologies would also significantly enhance the use of computers for education. The concept of "hypertext and hyperlinks" was not new. It was demonstrated at Stanford University by Doug Englebart in 1966, along with the first demonstration of the computer mouse. It was many years later in 2006 that I had the privilege of meeting Doug and having lunch with him at Berkeley University near San Francisco and seeing the old black and white video of that first demonstration of the mouse which became known as "The mother of all demonstrations". Even though he was well into his 80s, Doug still had a very enquiring mind and a twinkle in his eye. Our first experience of hyperlinks was in our original desktop publishing software package called Ventura. The use of hyperlinks and hypertext was a major advance for educational technologies because it allowed learners to explore related ideas in a new self-directed way but it would be sometime before these hyperlinks would be substantially overtaken by search engine technology.

As the computer graphics got more sophisticated and computers became able to display thousands of colours instead of just 4, 16 or 256 and at the higher resolutions of VGA and SVGA, the potential of computers for education became more apparent and an exponential growth in computer based education started. It was the birth of the internet in 1992 that created a whole new education possibility both through asynchronous education (not real time) and online synchronous education. Asynchronous or on-demand education was mankind's most basic use of technology to deliver what traditionally would have been on a printed page to a global audience and make it accessible 24 hours a day 7 days a week. The human intervention that can bring meaning and understanding to this content would be provided by either telephone calls or exchange of emails. My first introduction to asynchronous, collaborative learning was through a software package called Caucus which was a surprisingly functional (for the 1990s) online forum and knowledge management system. I was in a Community of Practice project led by a distinguished Academic called George Por in a project known as the Knowledge Ecology Network or "KEN".

I cannot recall now how I came to be involved, but I became part of an online community of about 100 participants discussing and debating the use of technology to generate knowledge. For about a year around 1998, I would often spend more than an hour a day sharing views and ideas with a wide spectrum of people from all over the world. Amongst those people was Nancy White from Seattle, a specialist in online communities of practice and whose online nickname was "Choco Nancy". All of our interactions were through text posts and the biggest influence it had on my life and career, apart from opening my eyes to its potential, was the expansion of my international network of contacts and my introduction to some very gifted and respected people. As a result of these interactions, I attended a conference in Seattle and stayed with Nancy and her family, something I was later able to reciprocate when Nancy came over to the UK to speak at a conference in Nottingham.

It was the Caucus software platform that gave me the ideas for the Community Commerce and Knowledge Network (Comknet) project that I secured funding for at the end of 1998. My enthusiasm, in retrospect, was at least in part because this was a novel application that I wanted to explore to the maximum. I remember being puzzled by the attitude of one of KEN's London participants when he displayed some scepticism about the ability of forums like this to engage people on a regular basis. I now know that these online forums can become "old hat" to experienced users and that the initial enthusiasm can wane very quickly without the skill of a good moderator / facilitator and/or without the basic motivation to participate because of a passion for a topic or an essential need for knowledge.

It wasn't until the second half of the 1990s that search engine technology began to emerge not only on the internet but also on desktop computers and corporate networks. One of these early search engine technologies that caught our eye was from a UK company called Autonomy based in Cambridge. It was our understanding that the technology behind Autonomy came out of research on fingerprint matching in which "pattern recognition" provided a much more powerful basis for search that than any previous text string matching. I had been using our RBase database software for such searches since the 1980s and this was well proven, especially when searching databases but text in documents is a totally different requirement because context is extremely important in achieving the most relevant results.

The techniques used by Autonomy were able, for one of the first times, to make connections between bits of information that previously only human beings were able to make. Their technology found its way into some early commercial internet applications and I seem to recall

British Telecom introducing an internet based facility that could connect people together who shared a common interest in something. It made these connections by tracking the web pages people viewed and showing a network chart with clusters of people around common topics. This kind of technology has been successfully used by both eCommerce and social networking sites.

It was also in the late 1990s that desktop video conferencing became a practical possibility and I settled on a system made by Intel and called "ProShare". This system required the speed and quality of an ISDN line that could reliably deliver speeds of up to 256 Kilobits per second. Video Conferencing was not new by this time. Even towards the end of the 1970s Post Office Telecomms had been pioneering boardroom technology that could operate over private circuits. Once ISDN became available widely in the UK, many corporates began investing in these boardroom solutions with a large television monitor and camera control unit to handle the video and sound. In these early days, like any "hot" new technology, the expectations of vast savings in travel costs were not really met and, in my view, this was entirely understandable because the experience of using boardroom video conferencing was inferior to both face to face meetings and, possibly more importantly, to the broadcast quality of video and sound found in the home. As a consequence many video conferencing systems were condemned to gather dust in boardroom cupboards.

Today, video conferencing over broadband networks is virtually capable of addressing these two basic perceived weaknesses, especially with systems like Cisco's Telepresence but from an educational and also a cost effectiveness perspective, it is not the ability of video conferencing to realistically mimic a face to face meeting, it is the raft of tools have been developed during the 21st century that can make these collaboration technologies more productive and cost effective than face to face learning.

Intel's ProShare was part of a family which included a boardroom system called Teamstation which was a powerful server computer customised with the video processing card, good quality teleconferencing microphone and video camera that could be remotely controlled. The key to ProShare and to the many solutions that are now commonplace was the ability not only to see and speak to the remote person, but to be able to share documents and collaborate, whether this was simply viewing a Powerpoint Presentation or jointly working on an office document. Whilst Proshare was a major advance in the consumerisation of desktop conferencing solutions, it should not be forgotten that I was able to deliver a presentation remotely to students at Kyoto University in the late 1990s using the very basic tools of Microsoft NetMeeting which, even in those days, had some of the facilities that have proved so popular with social networks and with Skype teleconferencing. With NetMeeting you could search a directory for contacts and send an instant message to them. This facility could have rebounded very badly on me during that Kyoto presentation because it was literally only about 15 minutes after my presentation, whilst I was still on line, that I got a video connection from a woman in Canada who was clearly interested in its use for non-academic purposes that would not have been appropriate for the Kyoto audience.

As the 1990s were drawing to a close, there were also some major developments in search engine technologies and knowledge management tools. These were in the days before Google became the dominant search engine force that it is today. I experimented with a number of these technologies both for internal and external use and also to gain experience that I could pass on to my clients. It was during the period 1998 to 2000 that I became actively involved in the virtual classroom and collaboration technologies that form such an important part of my life and career today. In many ways, it is my experience that, even after these facilities have been commercially available for over a decade, most organisations still do not fully understand the potential of these technologies as both educational and business development tools. It is perhaps ironic that, as in so many of the powerful technologies available to businesses, it is in their domestic use in social networking and Skype applications to friends and family that brings about the revelation of their potential for business.

It was the use of eVideo on my first live global virtual conference to over 300 online participants that led to my introduction to HP Virtual Classroom and on to the myriad of other virtual classroom, seminar and conferencing tools that I have used over the last decade. I believe that I have been something of a pioneer in the concept of "hybrid" events which bring together physical and virtual participants in the same space. It was this ability that attracted the interest of an academic called Ahmet Cakir who was an active and senior member of an international association for Ergonomics that was called WWDU (Work with Display Units). This name has been changed in recent years to something which better reflects their mission in improving global understanding of our interaction with digital technologies.

It was through Ahmet that I got introduced to the couple who have become very valued friends and collaborators over the years. Professor Halimahtun Khalid of the University of Sarawack in Malaysia was at the WWDU 2002 conference in Berchtesgaden, Germany (http://www.ergonomic-institute.eu/files/contents.pdf) along with her husband, Professor Martin Helander of Nanyang Technical University (NTU) in Singapore. Ahmet had corresponded with me for several weeks before the conference and I agreed to use the ComKnet project HP Virtual Classroom to deliver a "hybrid" workshop to both those physically in Berchtesgaden, a virtual audience logged on to the virtual classroom and even a radio audience listening to the HFM community radio station in Market Harborough. To get to Berchtesgaden I had to fly from London Stansted to Amsterdam where I changed planes to fly onwards to Salzburg from where I had to catch the train to Berchtesgaden, a beautiful Alpine town which had been a favourite of Adolf Hitler in the war. Hitler used to take his mistress Eva von Braun there to stay at the Eagle's Nest on top of the mountain.

The change of planes went smoothly in Amsterdam but when my

WWDU logo of Ergonomics Instuitute

plane landed in Salzburg and I was the last person standing at the luggage carousel I realised that something had gone wrong. I registered my missing luggage at the airport and caught the train to Berchtesgaden without any of the details of my hotel on my person. These were in my hold luggage and somewhere between Amsterdam and Salzburg. I seemed to recall that the first instruction in the hotel directions was "Turn left out of the station". After that, my memory was a blank sheet. I turned left and walked up the hill trying to find a hotel name that looked even vaguely familiar without success before I finally faced up to my potential embarrassment and walked into one of the hotels and asked if they knew where I was staying or at least where the WWDU conference was. Fortunately for me, the town had a tourist information system that held the records of all hotel guests and where they were staying. I arrived eventually at my room and walked out onto its balcony to witness the most beautiful sunset over the mountains. I telephoned my wife to tell her I had arrived safely and waxed lyrical about the sunset and the fabulous view from the balcony before concluding the conversation with the afterthought "Oh by the way, I've lost my luggage". My luggage did eventually turn up two days later minus any of the valuables I had stored in it including my digital camera and foreign currency.

This WWDU 2002 workshop was a great success in which the technology behaved itself perfectly and the delegates at the conference hotel were very impressed by what for most of them was their first sight of these technologies. I was able to meet Professor Khalid and Martin Helander at the conference and begin the discussions which led to an even more ambitious workshop at the next WWDU conference in Kuala Lumpur in which I also involved remote presenters delivering to the physical audience at WWDU. My years of using virtual classroom technologies in my facilitation work for HP and Agilent technologies taught me a great deal about the effectiveness of these technologies compared to face to face training and about how to achieve best results. Virtual classrooms do have a number of significant advantages over face to face education and these include:

- Automatic capture and archiving of training content for later on-line access

- Student to student chat without disturbing the presenter
- Question and answer sessions can be handled throughout by a separate question manager
- Automatic recording of all comments, questions and answers for later analysis

The most common weakness of virtual classrooms is the lack of emotional commitment to the sessions compared to training where the student has to travel to attend a face to face session – it is very easy to nip in and out of virtual classrooms without the absence being noticed. My top tips for the use of virtual classroom would be:

- Use dual presenter techniques like they use on television and radio news programmes – two different voices keep the interest level high
- Use teleconferencing alongside the virtual classroom for audio access to the session
- When using remote presenters in a hybrid session, have a host or moderator on stage for the live audience and use questions and answers to the remote presenter

The other application which ran alongside the work with Agilent Technologies virtual classroom facilitation was a knowledge management system called eRooms which was used to provide secure shared access to the documents used in the training sessions and also to post the session reports and recommendations for continuous improvement.

My involvement in the Knowledge Ecology Network project and work with both Comknet and the Harborough Community Learning Network projects led me into many different opportunities to present at international conferences and collaborate on Lifelong Learning projects such as the European Rural Development by Education (ERDE) project. I was honoured to be invited to be a keynote speaker at Moscow's first eLearn Expo conference in 2003 and this led to my ongoing participation every year since then and a valued relationship with the Moscow State University for Economics, Statistics and Informatics.

Sally Ann Moore (SAM), organiser of the first Moscow eLearn Expo Conference

Moscow's first eLearn Expo was organised by Sally Ann Moore (SAM) who, like myself, was born in Lincolnshire and went to school in Louth. SAM is probably the best-read person I have ever met with an almost encyclopaedic knowledge of science and philosophy. Her father was a distinguished engineer who encouraged SAM to follow a similar career path at University before she worked in the computer industry with DEC (Digital Equipment Corporation), one of the competitors in my time in IBM. She had also done high level consultancy work on change management and eLearning and lives in France on a charming farm near Annecy. She is a lifelong friend with whom I continue to collaborate and whose farm I visit on my way to and from my annual holiday in Provence.

It was at this first eLearn Expo conference that SAM introduced me to two other significant people in my involvement in learning technologies. The first of these was Vladimir Tikhomorov who was the Rector of MESI and launched the very first eLearn Expo as well as being a main keynote speaker. He is a very imposing figure both physically and intellectually and although our conversations have always been through an interpreter, we have become very good friends. I was invited as a guest to celebrate MESI's 75th anniversary and also to run a serious games masterclass at MESI for the students studying Informatics, one of MESI areas of speciality. Vladimir is the patriarch of the Tikhomorov

Vladimir Tikhomorov former Rector of Moscow State University (MESI

eLearning "dynasty" in which his daughter Elena runs one of Russia's most successful eLearning companies. One of the traditions which the Russians have at special dinners with international guests is for each guest to propose a toast and/or tell a story before everyone at the table has a shot of vodka. If your dinner has a substantial number of guests at the table, this can be a challenging process.

The other important person I met at this first eLearn Expo was Alla Khaikina who now runs her own conference and exhibition organising company in Moscow and who has become a good friend with whom I am always happy to collaborate in support of her events. Alla organises the exhibition space at Learning Technology conferences and I support her with connections to international e-learning suppliers and experts. It is interesting to observe how the focus for these Moscow events has

Alla Khaikina Moscow's leading conference organiser

shifted over the eight years that I have been visiting Moscow each May/June. In the early years of the conference, most of the attention was focused on Learning Management Systems (LMS) which keep track of student training records and progress. There was very little focus on the actual technologies used to deliver content or support educational practices. This has changed in recent years when there has been a growing interest in the more advanced technologies for developing and delivering learning content such as serious games and the immersive technologies that I discuss in Chapter Nine.

My trips to Moscow have not been without incident. On one trip my plane was about to take off from Heathrow when it turned back from the end of the runway to go back to the departure gate. I noticed a man slumped in his seat about 3 rows in front of me and he was carried off the plane unconscious before we were advised us that he had most likely been a drug carrier whose package had burst open in his stomach. In removing him from the plane, the emergency chutes were accidentally activated so we had a 3 hour wait on the plane whilst this was fixed. When the plane landed at Moscow airport, all the Russian passengers applauded which I took to be a sign of appreciation that we had actually arrived after all the delays. I learnt later that on Aeroflot or Transaero flights this is rather worryingly a standard practice and probably a legacy of old Soviet days when planes were less reliable. We arrived late at night and as the taxi drove through the suburbs I reflected on how dim the street lighting was and how little things had changed since my first visit in the Communist era of 1966. It wasn't until I arrived at my hotel that I realised that the dimness of the street lights was down to the tinted windows on the taxi.

The software tools for creating educational content that could be accessed either via CD-ROM or over the internet continued to improve. The on-line virtual learning environment called Moodle was invented by Martin Dougiamas as an open-source technology that is now used by many millions of people. I maintained my interest in end user tools for creating learning content and one of the early solutions I explored was a package called Opus developed by a Banbury Software company called Digital Workshop. It was this company that I had visited many years before to discuss reselling their "paint" software package called

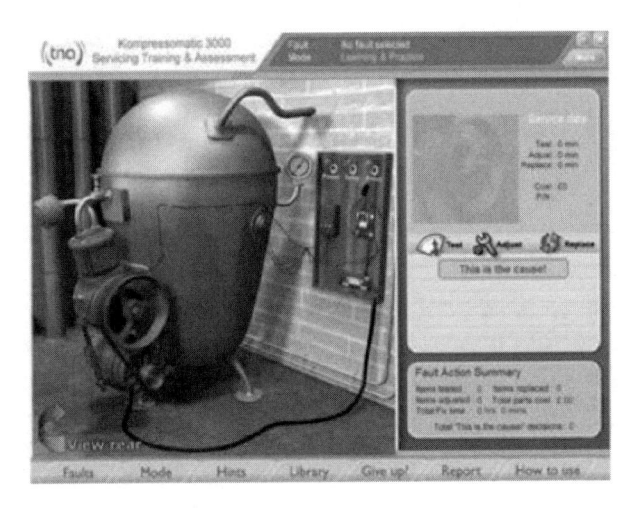

TNA Kompressomatic 3000 Service Training and Assessment Demonstrator

Paintshop which was later acquired by Corel to replace their own painting programme. Opus is an interactive multimedia development programme designed for use by training professionals and end users without the need for programming or graphics skills. This is a trend in software development that follows the pattern of Web 2.0 by empowering ordinary people to achieve extraordinary results.

At the very first eLearn Expo, one of my fellow UK presenters was Tim Neill, founder of TNA (http://www.tnanet.com/), elearning specialists based near Brighton, and winners of several elearning awards. Tim discussed and showed screenshots from some of his demonstrator and client applications and greatly impressed me with their innovative and highly creative approach. One of the first demonstrator applications he presented was designed to train and assess maintenance staff on fault finding and repair. Tim's Kompressomatic 3000 demonstrator shows an animated machine with simulated faults which the engineer can run diagnostics on and implement the necessary repairs before running the machine again. The innovative aspect of this application is that it does not simply award marks for the correct solution, it shows an animation after the chosen repair and calculates the cost in time and spare parts, giving a more real-world element to the training and assessment.

Most of TNA's work revolves around taking a client's existing

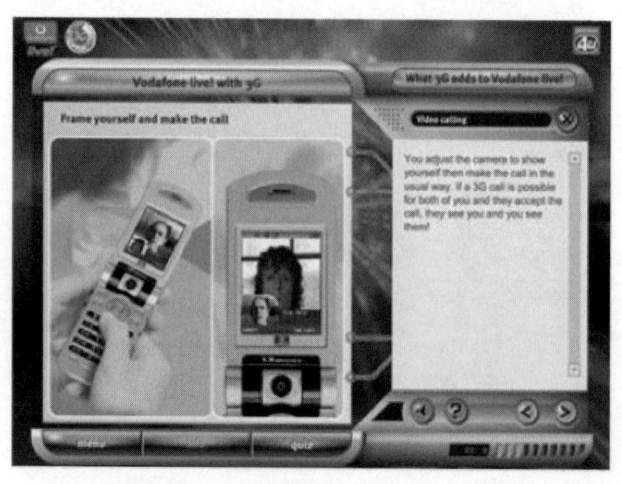

Phones 4U Retail Staff Training application

training solutions and transforming their cost effectiveness and ability to engage the workforce. One the TNA client applications discussed in Tim's presentation was the Phones 4U Retail training which not only has to be very dynamic and fast changing as new products come on stream, but also has to run on in-store technology and allow the client to manage their own learning content and capable of being updated in each store using the same telecoms infrastructure that these stores operate their normal overnight retail transactions with. Tim and I became friends and joined in the social activities organised in Moscow which included a visit to a typical Ukranian restaurant just across the road from the conference hotel. After our meal finished late in the evening, our group was standing waiting to cross the busy main road when one of the Russian exhibitor staff, Alice Akulova, stuck out her arm and stopped the first battered old car to come along and bartered with him for a taxi ride home. I was amazed that an attractive young woman would take such a risk in a big city late at night but as she explained the next day, and as I later found out to my cost, this practice can be considerably safer than using "official" taxis.

There have also been a couple of incidents in Moscow involving ill-health and an accident. On one occasion, Tim Neill was at the Moscow airport when he suffered bad heart palpitations and had to be taken to

Jacqueline Cawston of the Serious Games Institute with Cisco's Mike Morris at Moscow

hospital, much against his wishes. He fortunately made a swift recovery and was able to get back to the UK not too long afterwards but several years later in 2009, when I was Director of the Serious Games Institute, I took my Business Development Manager, Jacqueline Cawston with me to help with the Exhibition stand we had ordered for the show. It was Jacqueline's first time in Moscow so she asked me if she could stay an extra day to do some sightseeing. Because I had already been to Moscow several times and know my way round the main tourist areas I arranged to take her to the Kremlin and Red Square areas on the morning of my departure back to the UK so that she could orientate herself for her own sightseeing that afternoon and the next day.

We took the metro from the Mezdunarodnaya hotel to the Kremlin and as we were walking towards Red Square, Jacqueline tripped and fell forward onto her arms which she was holding up to prevent damage to the camera and bag she had in each hand. It was clear that she was in pain but she could move her fingers and there were no physical signs of damage to her arms so we purchased a scarf in a local store to use as a sling and continued looking round Red Square and the international boutiques which now occupy the former GUM department store that runs down one side of the square opposite Lenin's Mausoleum. She was still in some pain when we went to an "official" taxi stand outside Red Square and asked the driver what the fare would be for the 2 mile journey back to the hotel. The driver showed us a tariff table and pretended not to speak much English so we both got in and were driven to within about 100 yards of our hotel when he stopped the taxi and demanded 5000 roubles or 100 euros. We had to give him all our cash before he would let us out the taxi.

When we got back to the hotel, our main concern was to get some medical treatment for Jacqueline's arms so I left her in the care of the hotel management who called for a doctor. It was whilst I was in the Departure Lounge at Moscow airport that Jacqueline called from the hotel to tell me that she had been X-rayed and had broken both her elbows. By this time both her arms were in plaster from her hands to her armpits and she needed assistance in all her personal hygiene tasks. She was later taken to Moscow's American Hospital where she had to stay until a nurse could be flown out to bring her back to the UK. During the three to four days she was in this hospital she received some wonderful care and was even taken sightseeing in an ambulance by one of the staff and saw more of Moscow than she would have been able to do on her own. When Jacqueline got back to the UK she needed about six weeks off work to recover.

I had first met Jacqueline when I was employed as a Project Manager at De Montfort University (DMU) in Leicester from 2005 to 2006. She was in charge of graduate placement at DMU and every year organised a special employers fair and conference to act as a match making activity between employers and students. She impressed me enormously with her professionalism and with her organising abilities during that time and when I moved on to the SGI she was the one person I wanted most to join me to support the SGI's business development activities.

During the dark period in my career between 2002 and 2005 when I found myself doing night shift work at the Tesco Frozen Food warehouse, I still continued my interest and activities in elearning. On one occasion I finished my night shift, went home for a quick shower, and drove down to Heathrow to catch a flight to Moscow that same afternoon so that I could present at the eLearn Expo keynote session the following morning. I also tried to attend the major learning technology exhibitions, especially the BETT show held at Olympia in London every January.

The BETT show is the main international elearning conference and exhibition in the world, especially for schools, colleges and universities. Virtually every provider of elearning hardware and software for this sector of the market exhibits at this show. The BETT show is also an

The BETT Show at Olympia in January 2011

excellent venue to keep up to date with trends in elearning and educational technologies. There is a much smaller but equivalent show for the corporate market place around the same time of year.

It was in 2005 that De Montfort University gave me the opportunity to resurrect my career. I had been for an unsuccessful interview at Leicester College for a business development manager post and had almost given up hope at the age of 56 of finding employment that could not only give me a better salary but, more importantly, provide me with an opportunity to use my experience in a field that I have a passion for. The previous main interviewer at Leicester College had greeted me with the words "It's Disco Dave" when she saw me waiting in the reception area. It turned out that she had been at Birmingham University when I was running my mobile disco in 1970-71 but even this degree of familiarity with my talents had unsurprisingly proved to no avail on that occasion.

The Project Manager post advertised at DMU was a short term contract appointment for a minimum of one year to set up a Creative Industries Knowledge Network in partnership with the University of Derby and designed to build bridges between Higher Education and the Creative Industries. I went for my interview dressed as smartly as I could be and had an interview in the DMU Fletcher building with a panel which included Iona Cruickshank and Gerard Moran. I was encouraged by the fact that both Ioana and Gerard were close to my own age band and, as I drove back to my home in Lubenham I was praying that I would be successful. It was a wonderful feeling when I got a phone call the same day offering me the job.

I was keen to use my expertise in virtual classroom applications to support this work and, after doing some initial research, I chose Adobe Connect Professional as a platform not only to run virtual seminars but also to record both promotional and educational content to be available for viewing on a 24x7 basis. Fortunately the Faculty of Art and Design in which I was based had some very good video equipment and an editing suite that allowed me to record a voice and video track to support an introductory presentation about the Creative Industries Knowledge Network I was tasked with setting up and managing. I was based in the 7th floor in the tower block of the Fletcher Building and my time in the frozen food warehouse had improved my fitness so much that I was able to easily climb the stairs several times a day. My immediate boss was Iona Cruickshank whose office was in the Portland building about 100 yards away and her special interest was photography. Her holidays seemed to be spent researching and photographing drive in movies in Canada and the USA.

I was lucky enough to meet and work with some very interesting characters during my time at DMU including Michael Powell who ran the Game Art Design course and had strong contacts with the Games Industry, especially West Midlands based Codemasters and Blitz Games. Michael was a great believer in his students doing work with traditional drawing tools to create artwork on paper before they were let loose on the graphics workstation in his department space. It was Michael who also introduced me to the virtual world Second Life that later became an important part of my work at the Serious Games Institute.

De Montfort University were also involved in innovative research into 3D. Even in my time at Mass Mitec in the 1990s I had been shown a prototype of 3D television technology that did not require special glasses and it has been around 15 years before such screens became consumer items that are starting to appear in people's homes as 3D technology starts to take hold. Martin Richardson at DMU had done a lot of work on holographic photography and a holographic photo he had made of David Bowie held pride of place in his studio. At the time I was at DMU, he was also working on special holographic thread for garments as a way of preventing copyright infringement by fake garment manufacturers. I also worked closely with Dr Richard Hall who was the eLearning co-ordinator at DMU where the Virtual Learning Environment called "Blackboard" was in active use. Michael Powell began to explore the idea of letting the students on his course create their own eLearning environments and he pioneered the use of some of the emerging tools from Google.

The Creative Industries Knowledge Network project was at a time when technology was noticeably changing the nature and challenges of education. The internet and some of the collaboration tools that were becoming available began to change the role and status of the knowledge worker, not just in education but in almost all professions. I had already had first-hand experience of this phenomenon in 1999 when I met Frank Bingley, the self-taught creator of the Bigfern community web site in Market Harborough. Students coming into university to do digital media courses often had more experience of these applications than their tutors, and doctors were seeing patients come into their surgery with printouts from the internet covering their symptoms, diagnosis and treatment in a greater depth than a General Practitioner might be expected to have. At the seminars and workshops I ran at DMU I generally found that the creative industry practitioners, many of whom had lifestyle businesses based around a passionate interest, had a greater depth of both practical and theoretical knowledge.

In the fairly regular seminars and workshops that I ran, I tried to get interesting speakers to present from both industry and academia. On one occasion I ran a one day event at DMU which covered both mobile games and cinematic animation. I was very fortunate to get Dave Alex Riddett, the Director of Photography at Aardman Animations in Bristol, the world famous Oscar winning stop motion animation specialists. Dave was the son of Alec Riddett who designed and helped construct many of the sets at the Market Harborough Drama Society and it was Alec that helped to make the connection for me. By another coincidence, Alec had also worked with my boss Iona Cruickshank in a previous post. Amongst the industry partners I worked with was Channel Twenty Twenty, a video production company who were also pioneering internet TV and had an interest in launching a community TV channel over the internet. Channel Twenty Twenty was run by Rob Potter, a local businessman and social entrepreneur with a passion for promoting civic life in Leicester. Rob and his team played a very important role in the international conference that we organised at the Marriott Hotel in the early summer of 2006.

Multimedia technology was making continuous strides in both the performance and quality of desktop and laptop computing but also in the ease of use and functionality of the software, making it possible for the man in the street to achieve impressive results from domestic computers. I was involved in helping to organise and promote a special one day event called "Here, There and Elsewhere" at the old Phoenix Theatre in Leicester. The driving force behind this photographic and Multimedia production was a DMU lecturer, Lala Meredith Vula, who was an outstanding and controversial photographic artist whose mother was Scottish and estranged father was Serbian. The day, which I acted as Master of Ceremonies for, was a series of audio visual presentations about what it feels like to be a displaced person such as an asylum seeker or refugee. All the presenters used different techniques including video, data projection and multimedia to tell their own story of displacement but it was Lala's presentation which stuck in my memory.

Lala had not seen her father since she was a small child and had been brought up by her Scottish mother. She had never lost a curiosity about what her father's life was like in his Serbian homeland so she set out to meet him and use her considerable photographic skills to capture her story. During her presentation she showed many images of the poor and basic lives of ordinary people in Serbia using beautiful black and white photos of people and the artefacts of daily life such as the rear end of a motor car that had been adapted to be pulled by a horse like a cart. It was the combination of these fantastic images and her very moving, open and honest narration that left an indelible memory and has acted as a lesson for my own presentation techniques.

Towards the end of my time at DMU, I organised a one day workshop at the Leicester Depot, a city centre regeneration initiative that provide incubation, hot-desking, workspace and facilities for small creative businesses. The Leicester Depot was located in the heart of

Leicester's cultural quarter close to where the new Curve Theatre replacement for the Leicester Haymarket has now been built. This workshop was on games based learning and collaboration technologies and featured Martine Parry of the Apply Group and Kam Memarzia (now Kam Star) of the London based social game development company Playgen. The workshop also included an online live presentation from India by a games based learning company using virtual classroom technology.

Leicester Depot was a great example of a "creative cluster" in which small creative businesses operate from the same premises and have space in which they can communicate, share ideas and issues and collaborate. Apart from acting as an engine for innovation, this type of cluster can be very successful in building brand awareness for a city, town or village with its consequential attraction of investment and skills to an area. I took a lot of the lessons I learnt from working with the Leicester Depot Manager, Peter Chandler when I set up the Serious Games Institute (SGI) in the years that followed. Also at this workshop in Leicester was Kevin Corti who had founded a games-based learning company called PixeLearning who were later to become one of my first tenants at the SGI.

It was that session on Serious Games that perhaps helped me to secure the Director's post at the Serious Games Institute just a few short months later. My time at the SGI is covered in depth in Chapter Nine but education has been a very important application area for serious games and immersive technologies. I have spoken at many conferences on e-learning and education technologies and written quite a number of articles for publications such as Training Zone, Health Global and Defence Global. I will include a list of conferences and publications in a separate appendix but I would like to make some observations on the impact that all the technologies covered in this book have had on education.

Man's continuous quest to use machines and technology to improve productivity and effectiveness in all human activities has resulted in amazing cost/performance ratios in computing and telecommunications over the last 3 decades. From an education perspective this has made existing knowledge accessible on demand in rich and powerful formats from almost anywhere on the planet. However, it is not this accessibility of existing human knowledge which has been the focus of the real transformation in education in the 21st century, it is the technology developments in areas which encourage mankind's exploration and collaboration that are generating new knowledge at an unprecedented rate and changing the capacity and motivation to learn, and also the attitudes to learning in the new generations of young people.

Search engine and exploration capabilities developed by Google and others, not only for textual information but also for geographical and geospatial data, have made self-directed learning increasingly possible and in these processes the searches become more refined and more relevant to the individual. Not only this, but our searches, when combined with those made by other human beings, help the technology to begin to make connections between individuals and knowledge in ways which it has traditionally needed the intelligence of human beings to make. This ability to connect search patterns is not only used in education but also in commerce and in social networks where companies like Amazon have developed sophisticated techniques to try to both capture and retain customers by building "personal" relationships in much the same way that the local shopkeeper in the village community has always done with his customers throughout the last century.

It was this ability of e-retailers to personalise the internet retail experience that the Government's first E-Envoy, Alex Allan, referred to in a radio interview in 1999 when he evangelised about the potential of

Alex Allan former e-Envoy in 2000
Comknet Video

the internet for UK businesses. This personalisation in education can be described as "adaptive learning" in which artificial intelligence techniques are used to understand not only the capabilities of the learner but also their interests, passions and preferences. This adaptive learning technology will not only make it possible for learners to gain knowledge and skills through the devices they use every day, but will also go to inform e-portfolios for lifelong learning and assessment and will help to shape future career progression.

It is a common perception that very young children adapt to these new technologies very naturally and that baby boomers like myself can best learn how to use remote control devices in the home by asking their grand-children. These "digital natives", who have never known a world without technology, have early access to a whole spectrum of digital media and devices for the entertainment, learning and socialising. There is evidence that such familiarity with multiple media devices is leading to a quantum shift, not only in mankind's ability to multi-task or "task swap", but also in the brain structures of our young people and "hard-wiring" such multi-tasking capabilities into their brain structures.

I have written partly "tongue-in-cheek" articles which reflect on these possible changes by using the physical differences between the male and female brain that have evolved over the centuries to question what the implications of these changes might be. It does seem to be accepted wisdom that the male brain has evolved to major on an ability to focus or concentrate on a task to the exclusion of everything else. This ability to focus such that, in periods of high concentration, a man's ability to hear is affected, is almost certainly a legacy of the days of the caveman when the most successful hunters were those who could focus on killing the prey that earned their survival. A woman's brain may well have been shaped by the female role of socialising and "gathering" which, over centuries, has made the female brain more able to multi-task than the male brain. Could it be that in the 21st century when the traditional roles and males and females are becoming blurred that technology is homogenising the human brain of both boys and girls to make both sexes equally capable of multi-tasking, and will these capabilities be developed at the cost of losing the ability to concentrate and focus for extended periods.

All these changes in educational technology are having a severe impact on the role of the teacher whose traditional practices of "chalk and talk" and being "the sage on the stage" are becoming redundant as today's students are more demanding and also on many occasions more knowledgeable than their tutors. These trends require teachers and educators to become more like coaches or mentors whose role is to facilitate and direct students to make the connections they need to pursue their self-directed learning practices. Mobile technologies have also had a big impact on education by making learning an activity which can be accessed anytime from anywhere on devices like mobile phones and the new flat screen "iPad" devices.

Education technologies have come a long way since I sat in rapt attention at Staniland Primary school in Boston listening to Mrs Derbyshire exciting our interest with Biggles' adventure stories. The BBC's Listen with Mother programmes of my early days have been replaced with byte size chunks of learning that can be downloaded onto a mobile phone that we can access on a bus journey into work. Although we are only just at the dawn of a new era in which the machines that help us to learn will use their collective intelligence and senses to get the best out of our human potential, the pace of change is accelerating and the technologies covered in the next two chapters will be at the forefront of what we can expect in the next few years.

CHAPTER EIGHT
See Me, Feel Me, Touch Me, Heal Me

One of the most distinguishing advantages humans have over machines is the range of senses that we have and our ability to process this information, make "intelligent" judgements and act on those judgements. Our sight, hearing, touch, smell and taste capabilities provide our brain with key measurement and assessment data that we use to make sense of the world around us and that we add to our memory and processing capabilities. We human beings have a desire and an ability to understand, manage and control the multi-dimensional environment in which we all live. We are conscious of the dimensions of time, space and state and it is our ability to measure these dimensions and analyse their properties and behaviours that enables us to learn about the past and predict the future. The constraints that these dimensions impose on our lives are both a comfort and a frustration. We are comforted by the fact that our knowledge of the properties allows us to plan and manage our daily lives yet frustrated that throughout mankind's history we have had a very limited ability to conquer them. This chapter is about the technologies that have been developed in my lifetime to increase our

Where am I?

ability to understand and therefore master our environment and about the intelligence we are transferring to machines that may yet make that aspiration impossible.

My father was a carpenter, joiner and undertaker and the tools of his trade included "technology" that allowed him to take the measurements that assisted his skills, experience and intelligence to craft wood to the specific requirements of any job, be it making window frames, book cases or coffins. When he worked as an RAF Ground Crew Technician in India in the war, the Lancaster Bomber aircraft were equipped with technology that enabled pilots to navigate and fly on missions but, as in my father's woodworking trade, much of the success of these missions had to be attributed to the human skills and experience of the pilot and his ability to harness all his senses to interpret the data and translate it into control of his aircraft. Apart from the tools of my father's trade, the only other measurement technologies in our Wyberton household were in our kitchen to help my mum get recipes right, the living room where our one clock frequently prompted my dad to yell "Look at the time, look at the time!" as he raced out of the door to the bicycle which would carry him to work, and in the hallway near the front door where the barometer, after some tapping of its glass panel, would tell us the temperature and air pressure and very little else.

Like all young children, I would explore this magical world around me and over time would learn about the physical properties and behaviours of the objects and living things I encountered and my familiarity, understanding and ability to predict the behaviours of these would grow. With this growing familiarity and understanding of the world there was also a change in my relationship with not only objects but other living beings I encountered. I came to know them and my attitude towards them was shaped by my experiences. I began to "love" my home, my family and my familiar toys and my response to different people and objects was shaped by my relationship with them. I learnt how to manipulate objects to achieve my objectives and my behaviour and reactions were shaped not just by the physical properties of those objects but by my relationship with them and how I "felt about" them.

This combination of understanding, experience, familiarity and relationship with objects, people and places is refined and made more sensitive by our senses. My uncle knew from the sound and feel of his motor car, "The Bomb", how the car would behave and consequently when he needed to drive it more gently. We human beings also have this kind of relationship with physical places and this affects our emotions and shapes our feelings and responses to them. These capabilities, emotions and behaviours were, in my early lifetime, very much the domain of humans and not found in machines.

The human brain has an incredible ability to process the data that our senses constantly bombard it with and to compare and contrast all this data with previous experiences, apparently ignore what it considers unimportant, and register and act on what it considers important. It is this registering of specific sensory information that can capture and recall on demand sights, sounds, smells, tastes and feelings from years gone by triggered by a word, a sound or a smell. There are moments in my life when I can conjure up a picture in my mind when I hear a song, or almost smell the Sunday roast when I hear the words "Two way family favourites", a radio programme that always seemed to coincide with Sunday lunchtime. It is this highly individual and personalised reaction to specific situations or environments that has distinguished us human beings from machines throughout history.

As we go through life, we have an almost paradoxical need for simultaneous predictability and unpredictability. It is predictability that brings comfort and constancy, but predictability is the enemy of learning and development and so our minds crave for those uncertainties that bring challenge and excitement to our lives. It is as if our brains demand this balance between familiarity and the unknown to be at their most effective. Too little familiarity leads to stress and overload and too much familiarity leads to boredom and contempt. Could this provide us with a better understanding of how the human brain has evolved differently for men and women and how the challenges of 21st century life with its accelerating pace of change will determine its future evolution?

There are several instants in my life that I can recall with absolute clarity yet can provide no rational explanation as to why, at these moments, I felt totally content to the extent that if my life had been extinguished at those precise moments I would have been happy and have felt no loss for anything unfinished. One of these occasions was when I was about 14 years old and was in the Boy Scouts. I had already been in a number of the "Boston Gang Shows" under the direction of the much feared and respected Sid Burgess when we had gone down by bus to Golders Green in London to watch the famous London Gang Show directed by the legendary Ralph Reader. I remember nothing at all about the show or the journey but I do remember being dropped off the bus and walking home down Tytton Lane East after midnight on a crisp starlit night and hearing the solitary sound of my footsteps crunching on the frosty pavement. I looked at all the majesty of the heavenly stars so bright and clear on the cloudless night and I became one with the universe. God was in his heaven and all was well with the world. As these emotions come back to me and tears start to well in my eyes at the memory I can hear the Steve Miller Band playing "Quicksilver Girl".

Our senses are our primary interface to the physical world and it is our brains' interpretation of the data provided by our senses that triggers our actions. Machines communicate with us by delivering messages to our senses and although today we are now able to communicate back to machines in a whole variety of different ways, it is in the development of the technologies that machines use to understand and interpret our communications that has undergone the most radical transformation. For most of my life, machines have had very little and/or very specialised sensory capabilities and those have been used to trigger pre-programmed responses such as thermostat controlled switches, timers or movement sensors that activate lighting or CCTV cameras. My life began in an era where human beings had sophisticated processes to interpret and act on information provided by machines but had only very basic tools to control them such as levers, switches, pedals, knobs and wheels. Today machines are able to understand and act on our motions, our voices, our presence and even our brainwaves. This chapter is about my experience of the journey that technology and machines have made towards matching and even exceeding our human sensory and interpretive capabilities.

One of the other distinguishing features of human beings (and other living creatures) is our awareness of the dimensions of space and time and the influence these have on our behaviour. Mankind has developed technology over centuries to support our awareness and understanding

SEE ME, FEEL ME, TOUCH ME, HEAL ME

of these dimensions and has used this understanding to build the infrastructure and society we experience from the time we are born. This growing understanding and awareness of space and time supplements our inbuilt genetic "programming" and humans have long sought to create technology which can help break down the barriers of space and time and extend still further our influence, control and ability to shape our environment. At the time I was born, mankind had not yet entered space and, for the vast majority of people on the planet, there was little or no possibility of directly influencing anyone or anything outside of their immediate location. Only those who were in control of broadcasting and publishing had the powers to be able to do this. The world was a largely disconnected society in which the future was shaped and made predictable by the past.

Like all children of my generation, I settled into a routine that had been shaped by history, tradition and location. I became aware of the rules and patterns of my life and the constraints that space and time imposed on me, consciously and unconsciously organising my actions to observe these constraints. I learnt how fast I could walk or run and to predict how long it would take me to cover a certain distance. This knowledge of space and time programmed my actions so that I would allow ten minutes to get from my home to the bus stop to catch the bus that the timetable and/or my previous experience told me would be passing at a certain time. In our house I had a world atlas and a globe that gave me a perspective on distant lands that my imagination, coupled with what I might be able to read or see in my encyclopaedia, would paint pictures of. When I travelled on the train with my grandparents to new places, my senses and my memory would capture massive amounts of data that would in future times enable me to "know" where I was and to act according to that knowledge of space and time. In the 1950s, the technology did not exist to enable machines to have this awareness of location and it was only clocks and timers that could be used to programme machines to perform functions at set times.

Human to human communication has evolved over millions of years and is a very sophisticated process that is a highly personalised mixture of genetic programming and individual experience. We learn how to interpret signals from other human beings as well as how to signal to other human beings what our needs and/or desires are. All human beings have a high degree of freedom of choice on how to react to these signals and our reactions tend to be based on both interpretation of those signals and an understanding of the consequences of our actions. Our human need for this complex combination of predictability and unpredictability means that we do not always respond in the same way to the same set of circumstances. We are programmed to a greater or lesser extent to take risks motivated by a desire to explore or the excitement of the unknown or unpredictable. The technology that existed in my early life was not designed to be unpredictable, with the possible exception of ERNIE the random number generator that would each month select premium bond numbers for cash prizes, the latter day combination of a savings scheme and the national lottery.

The constraints of space and time and state were often a frustration in my childhood days. Like everyone else I looked forward to future events and often wished time would go faster. The car journeys to the seaside at Hunstanton seemed interminable and we used to pass the time by singing songs, playing "I-Spy" or looking out the car window to see if we could see the sea. Occasionally we would look for a white horse in the fields and perform the ritual chant "White Horse, White Horse bring me good luck" three times. The technology we used to transport us to these much anticipated new destinations, whether it was bicycle, car or train, had no intelligence that allowed it to "know" where it was or change its behaviour based on that location.

In these times there was no technology that could recognise an individual human being and use that information to change the behaviour of a machine or object. Human beings possessed these kinds of powers of recognition of time, space and state but machines did not. These limited capabilities of technology and machines in the 1950s and 1960s placed a dependence on the senses, intelligence, memory and skills of other human beings to make these measurements and act accordingly. Because of these limited machine capabilities, jobs for human beings to measure, record and act on this kind of information were an essential part of daily life.

As the 20th century advanced to its close, mankind began to make substantial strides in breaking down the barriers and time, space and state using the technologies that have been described in previous chapters. Transportation technology enabled me to experience Berlin, Moscow, Leningrad, Helsinki and Copenhagen in 1966 and in the early 1970s brought the joy of mountain scenery and skiing into my life. Telecommunications gave me a career and enabled me to influence other people in distant locations by phone, telex, fax, diskfax, the internet, virtual classrooms and video conferencing. Computing technology enabled me to manage my finances, create presentations and plan for the future. Imaging technology has allowed me to capture photographs and make these accessible in electronic format. Through entertainment technologies I have been able to bring enjoyment and understanding to people I may never meet in person and education technologies have both brought me knowledge and understanding and have empowered me to inform and influence others.

It is only in the last quarter of the 20th century that we began to develop technologies which might be described as transferring our intelligence to machines to enable them to not only communicate more effectively with us but also help them become more conscious of the dimensions of space, time and state and thereby manage our world for us.

One of my first recollections of a technology which enabled a machine to automatically recognise me was the use of a plastic card with a magnetic stripe. I cannot recall the exact time when I first used an ATM (Automated Teller Machine) to obtain cash from my bank but the first ATMs as we now know them began to appear in the USA in 1971 (http://www.atm24.com/NewsSection/Industry%20News/Timeline%20-%20The%20ATM%20History.aspx) and it wasn't until 1974 that the connected on-line ATMs began to be installed. In my IBM days, I used to use the ATM as an example of how Information Technology could be used to deliver competitive advantage by providing innovative customer services that were a quantum advance on what was previously available. ATMs are a classic example of this because their introduction in the "hole in the wall" suddenly made cash available on a 24x7x365 basis. The ATM machine "recognises" the individual customer through the magnetic stripe on the bank card and the customer verifies their identity through a Personal Identification Number (PIN) code. This technology allows us to have a personal relationship with a machine in which the technology conducts transactions that are individual to us. Of course there are always human beings who seek to defraud such human-machine personal relationships by fooling the technology into believing a false identity but human beings throughout history have performed similar frauds on other human beings by impersonating or falsely representing another person.

The card with the magnetic stripe over the last 40 years has become a commonplace human-machine communications interface that opens up the possibility of personalised relationship with not only other machines, but also physical spaces. The magnetic information which machines can read and understand (but humans cannot) changes the dynamics of the human – machine interaction and enables machines and physical spaces to respond selectively and "intelligently" to the individual. Security pass cards can have these magnetic strips programmed to allow specific access to rooms and technology on an individual to individual basis and begins to mimic human to human interaction in which our interaction with other humans is determined by our relationship with them and our knowledge of them. Though these early processes were flawed and not foolproof they were the beginning of the interface and sensor technologies that today and tomorrow will transform our relationships not just with machines and spaces but also with other human beings as we continue this seemingly irreversible process of transferring our intelligence and knowledge onto "the machine".

Today we use cards with magnetic stripes or "chips" to identify ourselves and communicate with all kinds of machines, principally as a mechanism for commercial transactions, which clearly necessitate proof of identity, but this increasing focus on electronic transactions whether via the internet or in a retail outlet is also massively influencing the personal relationship we have with machines and technology because these transactions are tracked, analysed and used to give technology a better understanding of our behaviours, likes, preferences and all kinds of other attributes of our lives and personalities. This is equivalent to the process we go through as human beings when we meet with other people and get to know them better. The more intimate our relationships are and the more frequently we meet, the more we share with each other and the

World Summit on Innovation and Enterprise in Oman 2006

better able we are to interact in meaningful and effective ways. The sensor, interface and location technologies we have developed in the 21st century are not only being used to enable machines to treat us in a more personal way, such as the techniques used by Amazon and other E-Commerce specialists, but they are also aggregating and analysing millions of such interactions to enable machines to better predict the likes and preferences of other humans like us and respond to us in the same way we do with our friends when we make suggestions to them based on our relationship with them e.g. "You really should read this book – I know you will like it!".

The implications and significance of machines being aware of the dimensions of time, space and state and being able to respond in an intelligent and personal way were brought home to me on a beach in the Middle East. In 2006, largely as a result of my previous contact with Debra Amidon of Entovation as part of the HCLN radio chat show on Innovation, I was invited to Oman to participate in the first World Summit on Innovation and Enterprise, a lavish conference held at the Shangri-La's Barr Al Jissah Resort near Muscat. It was a wonderful conference held in a luxury brand new five star resort setting with its own beach and one of the most memorable parts of this conference was a breakfast workshop delivered by Deepak Chopra, the world famous visionary and sage. This very informal session took place on the beach and Deepak's talk and discussion focused on the theme of synchronicity and serendipity. One of Deepak's most relevant quotes in this context was:

> *"According to Vedanta, there are only two symptoms of enlightenment, just two indications that a transformation is taking place within you toward a higher consciousness. The*

first symptom is that you stop worrying. Things don't bother you anymore. You become light-hearted and full of joy. The second symptom is that you encounter more and more meaningful coincidences in your life, more and more synchronicities. And this accelerates to the point where you actually experience the miraculous. (quoted by Carol Lynn Pearson in Consider the Butterfly*)"*

Deepak talked about his professional experience in neurosciences and also about the concept of parallel universes in which the physical world of which we are fully conscious is overlayed by an "energy" universe that we are not conscious of. He spoke about the phenomenon of simultaneous and spontaneous creation and innovation where ideas and inventions are developed at exactly the same point in time without any apparent physical connection between them. Experiments have been done on this "telepathic" kind of communication not only between humans and humans but also between humans and animals. Anecdotally, I have had similar experiences with my pets who seem to know exactly when I am coming home even when my trips and times have been very irregular. Deepak used this synchronicity to offer an insight into our own experiences of what humans call serendipity – those occasions when you bump into someone unexpectedly at exactly the right time and place when the probability of doing so is extremely remote. People who experience serendipity on a regular basis (I count myself in this category) often describe themselves as being very lucky but Deepak argued that it wasn't so much luck as being open and able to recognise such serendipitous events and exploit them. We use our senses, intelligence and consciousness of time, place and connections to recognise opportunities when we see them. Deepak argued that these serendipitous events happen to everyone with the same frequency but most people are not open to such events and consequently don't recognise them. Now, in the 21st century, machines and technology are evolving to the point where they are as able if not better able than human beings to recognise and exploit this phenomenon.

It has only been in the latter half of my career that I have found myself involved in disciplines and amongst experts whose work has focused on the kind of measurement and interface technologies which are bringing consciousness of time, space and state to machines and which are vitally significant and important to the future of mankind. This part of my career, which only really began with the Community Commerce and Knowledge Network Project that I devised and secured funding for in 1998, has three separate but highly interlinked strands.

The first of these strands is through the Global Society Dialogue initiative which is largely based on understanding and managing the impact of Information Communications Technology (ICT) on sustainable development. It is through this strand that I have worked with visionary and passionate leaders such as Professor Franz Josef Radermacher and Dr Thomas Schauer.

The second of these strands is in the field of ergonomics and human computer interfaces (HCI) where I have been fortunate enough to have met Professor Halimahtun Khalid of Damai Sciences in Kuala Lumpur who, along with her husband Professor Martin Helander, is not only internationally recognised and respected as an expert in this field but is also passionate about the potential of ergonomics to build a more sustainable and equitable future for mankind.

The third of these strands is through the International Society for Digital Earth (ISDE) whose members are specialists in the sensor and location based technologies that help us to understand and manage the space, time and state dimensions of planet Earth and who are committed to using their growing knowledge to tackle some of the most important and urgent environmental and societal challenges of our time. ISDE has brought me into contact with people such as Professor Milan Konecny, Dr Tim Foresman, Professor Manfred Ehlers, Dr Temenoujka Bandrova and Richard Simpson, all of whom have played and are playing pivotal roles in shaping the technologies which will determine our future.

Global Society Dialogue Strand – ICT and Sustainable Development

Towards the end of the 1990s, I had joined an organisation called Computer Professionals for Social Responsibility (CPSR) indirectly

Professor Franz Josef Radermacher, Global Society Dialogue

through a meeting at a conference in Seattle with Doug Schuler who was on the CPSR Advisory Committee. The CPSR was involved with a new organisation set up in 1998 and called "Internet Corporation for Assigned Names and Numbers" (ICANN) and it was through this connection that I volunteered to be involved in a European Chapter of CPSR. This involvement led me to be invited to the first meeting of the Global Society Dialogue (GSD) held in at the University of Ulm's Research Centre at Reisensburg Castle in Germany.

It was this meeting of GSD that introduced me to some very key people and ideas about some of the serious issues facing our future sustainability and the role that technology is playing in both creating the problems and providing the solutions. The GSD meeting in that beautiful castle setting brought together an eclectic mix of people and

Schloss Reisensburg venue of first Global Society Dialogue Meeting

organisations. There were academics like Professor Radermacher and Professor Milan Konecny from a wide range of disciplines, representatives like Milda Hedblom and Jyoti Parikh from non-Governmental Agencies, small business entrepreneurs like myself and individuals like Lachhe Bahadur, a schoolteacher from Nepal and Stella Tabirtsa who worked for a media organisation in Moldova. We came from all over the world and had a wide spectrum of life experiences between us.

In 1992 Al Gore had written a book called "Earth in the Balance : Ecology and the Human Spirit" which talked about a "Global Marshall Plan" to harness the earth's finite resources and build a sustainable and equitable future. This concept was further developed by Professor Radermacher who was the main driving force behind GSD. Franz Josef has a passion for trying to make a difference to the world we live in and a rare understanding of the impact of technology on our globalised consumer society and the implications for a planet with limited finite resources.

Together with Doctor Thomas Schauer who was also based with Professor Radermacher at the University in Ulm, Professor Radermacher wrote the book "The Global Marshall Plan" and even penned a musical

Challenge of the Digital Divide – Collective Views of GSD Project members

in collaboration with the singer Solvig Wehsener about the future of a world dependant on dwindling supplies of Oil and called "You can Believe". This musical was written before the first invasion of Iraq and was uncannily accurate in its predictions. The basic premise behind Professor Radermacher's work was that our consumer society, based on disposable goods and unending periods of growth, is unsustainable in a world of finite resources and that the only way forward is to plan for lower levels of growth in the developed world to allow the developing nations the higher levels of growth needed to bring a more equitable balance to the Global Society. I had the privilege of helping Professor Radermacher and Doctor Schauer to check the English translation of their book.

The GSD project funded a number of meetings and public workshops as well as the publication of a series of papers written by GSD members. These papers were presented at a Public conference / workshop in Vienna in 2001. In my own paper entitled "Universal Communication: Power to the People or Pathway to Destruction" I forecast the development of mobile communications and its potential impact and wrote:

> *"Today's mobile communications technology is fundamentally different than any previous quantum leap, because it places the power to communicate in the hands of the individual citizen, and shifts the balance between consumer and producer. We are moving towards a situation where technology will empower individual citizens to communicate with anyone, anything, anytime, and anywhere. There is already strong evidence that the effects on society by such individual empowerment can also be a threat to sustainability and stability. The disenfranchised individuals in society now have the tools to cause dramatic, global, and unpredictable impact; and terrorist networks can easily use the new infrastructure as well."*
>
> *"In summary, the new communication technologies are presenting challenges to a sustainable global society. We have to find a new balance between citizen empowerment and social responsibility; value and costs; culture of Interdependence and individual independence; consumption*

and preservation of resources; and free market and intervention. We will have to discuss, on the one hand, access to communications as a fundamental right, and on the other hand, control mechanisms like netizen unique ID or positioning systems."

This prediction about the development and growth of the mobile internet and communications has proved to be all too accurate and in August 2011 there were large scale riots and looting in major UK cities activated and made possible by mobile communications and social network applications. The "sensor" networks that have been put in place since the time of this paper in September 2001 such as CCTV cameras, social networking archives, GPS tracking in mobile phones and text message monitoring all helped to identify many people involved in these riots and bring them to justice but the experience of these riots also highlighted the problems caused by the text encryption technology embedded into Blackberry mobile phones which made it difficult if not impossible to identify some of the leaders of the rioting and looting. This should act as a warning to society about the need for proper balance between citizen rights and responsibilities and the importance to a sustainable future of being able to link cause and effect.

Another article in the same publication was written by Maja Gopel of YOIS (Youth for Inter-Generational Justice and Sustainability) and titled "Too old for Surfing? The Generation Gap in the Information Society." Maja wrote:

"Although retired people do not need computer skills for work anymore, these skills can enable contacts and facilitate integration into social life. Since older people stay home a lot more, and lack the personal contacts aligning the participation in working life, often media can become a substitute for them. As reports show, seniors with Internet access stated that learning how to use a computer strengthened relationships with friends and family, stimulated their mind, and gave them something in common with younger people."

> *"Seniornet, an internet platform for people older than 50, asked its users how they learned to use the Internet. Many got help from relatives, peers, or took classes and at the same time taught themselves. There need to be initiatives to offer guided first steps with the Internet. These have to be actively brought into the lives of older people, by promoting them through the media and organisations which are usually in touch with older people. Luckily, this is happening more: There are Senior Internet Cafes and Learning Centers; networks offer support with toolkits for new initiatives; and the integration of computers and the Internet in retirement homes is being seen as an enrichment.*
>
> *At the same time, steps are being taken to adjust the design of hardware and content, in order to meet physical challenges (e.g., sight, motion) older people may have. There are predictions that there will be a breakthrough of the Internet in all segments of society, when handling it will be as easy as using a TV remote control."*

The people involved in the Global Society Dialogue (GSD) were an outstanding collection of human beings from whom I learnt a great deal, not just about the challenges faced by our global society, but about humanity and our potential to make the world a better place for future generations. We had several meetings of GSD in different European locations and developed strong bonds between us that have led to frequent communication and collaboration. It was through these connections and collaborations, combined with my own use of communications and knowledge management technologies that I became involved in the Human Computer Interface and Digital Earth strands.

Ergonomics and Human Computer Interface Technology Strand

One of the phenomena of the 21st century is the emergence of the "silver surfer", the older person who is using the extra discretionary time and money brought about by retirement to explore the use of technology to enrich their lives through lifelong learning and a wider network empowered by not only better communications but improvements in interface devices and artificial intelligence applications. This highlights another current serious social and economic issue, the "Ageing Society" in which life expectancy is rapidly increasing which, coupled with lower birth rates in the developed world, means that a smaller working population is likely to have to fund an increasingly large retired population. The technologies discussed in this

Professor Martin Hellander and Professor Halimahtun Khalid

chapter such as sensors, location tracking and new interfaces have been identified as being key to tackling the ageing society issues by being able to better monitor elderly people, extend their independent living and deliver social care at a far lower cost.

In 2005, whilst I was working as a Project Manager at De Montfort University, I collaborated with Professor Halimahtun Khalid of Damai Sciences in Malaysia, whom I had first met at WWDU in Germany in 2002, on a project proposal which took a radically different approach to using technology to tackle the ageing society. Together with other partners we bid for funds under the European Framework Six (FP6) to develop a project we call "MIRACLE". This project focused on the positive aspects of the ageing society of the better health, extended life expectancy, life experiences and desire of older people to continue to be of value to society. Instead of proposing to use technology to effectively reduce the cost of caring for older people, MIRACLE sought to empower the silver surfer generation by making it possible not only for them to live independently but also use their experiences during retirement on paid activities that contributed to rather than cost society.

The Human Computer Interface is an essential part of our ability to communicate with machines in a meaningful way. Because the computer technology which today is embedded in and controls most machines only "understands" digital information represented by the 1s and 0s of bits there has to be a two way translation process to enable humans and machines to talk to each other. At the time I was born, human beings were only just beginning to understand how to instruct computers and even when I went to university in 1971, my handwritten Fortran computer programs had to be converted to punched cards with holes that were detected by a card reader before the computer could understand what I was trying to do. In these days there were very few visual display units or monitors and the computers mostly communicated back to humans through printed sheets of paper. Prior to the invention of the computer, machines communicated their "state" visually through meters with dials and gauges that measured physical parameters such as temperature, speed, pressure and so on.

By the time I joined IBM in 1979, we were able to communicate instructions to computers via a keyboard and they were able to communicate back via a small monochrome monitor with text information displayed in bright green letters on a dark green background. There were little or no forms of graphical communication and both humans and machines relied on a text format as the primary connection between them. In these times, human instructions had to be in a very precise format for machines to make any sense of them and machines were basically unintelligent slaves that could only do what they were told and only then if the instructions were presented in a format which the machine could understand.

In 1968, Douglas Engelbart did the world's first demonstration of a computer mouse but it was only when the Apple desktop computers began to appear did this new way of communicating with machines begin to become commonplace. Over the years since 1968, there have been really substantial developments in interface and sensor technologies that enable computers to understand and interpret the dimensions of time, space and state in ways which human beings have been able to do since homo sapiens first walked the earth. These interface technologies have emerged as a result of our growing demands to have relationships with machines which are as close as possible to what we experience with human to human communications. In the same way, over the last 30 years, we have empowered machines to communicate back to us through all our human sensory channels, even including a machine which can "print" food whose taste matches the "printout".

Digital Earth Strand – Sensor and Location Technologies

Digital Earth is the name given to a visionary concept by former US vice president Al Gore in 1998, describing a virtual representation of the Earth that is spatially referenced and interconnected with the world's digital knowledge archives. In a speech prepared for the California Science Centre in Los Angeles on January 31, 1998, Al Gore described a digital future where schoolchildren – indeed all the world's citizens – could interact with a computer-generated three-dimensional spinning virtual globe and access vast amounts of scientific and cultural information to help them understand the Earth and its human activities. The greater part of this knowledge store would be free to all via the Internet, however a commercial marketplace of related products and services was envisioned to co-exist, in part in order to support the expensive infrastructure such a system would require. It was this Digital Earth concept that was to become such an important part of my life in the 21st century and that would spawn many of the developments in the technologies that are the subject of this chapter.

It was however as long ago as 1968 when an American, Buckminster "Bucky" Fuller, first used the phrase "Spaceship Earth" in a book entitled "Operating Manual for Spaceship Earth" to reflect on some of the issues our global society is now facing with increasing urgency. He said *"...we can make all of humanity successful through science's world-engulfing industrial evolution provided that we are not so foolish as to continue to exhaust in a split second of astronomical*

Professor Milan Konecny, President of ISDE

history the orderly energy savings of billions of years' energy conservation aboard our Spaceship Earth. These energy savings have been put into our Spaceship's life-regeneration-guaranteeing bank account for use only in self-starter functions." Bucky was one of the first people to conceive of a planet as a living organism that could communicate with and be viewed by human beings on a holistic basis.

The strongest of my Digital Earth relationships has been with Professor Milan Konecny of Masaryk University in Brno, Czech Republic, whom I met as a fellow founder member of the Global Society Dialogue. He is a leading world authority on Geographical Information Systems (GIS) and a former President of the International Symposium on Digital Earth. It was my friendship with Milan Konecny that led to an invitation to present at the 3rd International Symposium on Digital Earth in Brno in 2003. It was at this conference in one of the breakout sessions that I spoke about my experiences in Community Informatics projects in the ComKnet (Community Commerce and Knowledge Network) and HCLN (Harborough Community Learning Network) projects.

Milan knows how to help others enjoy themselves through social activities rooted in cultural tradition and it was through the Wine Cellar evening at the ISDE conference that I got introduced to another future lifelong friend and outstanding human being, Dr Tim Foresman, who was to become a future Program Director of ISDE and organiser of the

5th ISDE conference that took place in Berkeley University in California in June 2007. It was at the ISDE conference in Brno that I first became aware of Remote Sensing applications and the mission of Digital Earth to create a "living model" of our planet that would help us to understand and better manage our environment. It is only in the 21st century that we have begun to realise Buckminster Fuller's vision of Spaceship Earth as a holistic entity that can communicate with us through sensor technologies and visualisation techniques that can show us a good visual representation of the health of our planet in all its diversity.

I was unfortunately unable to afford the cost of attending the following Digital Earth Symposium in Japan but I remained in electronic touch with my main contacts and by 2006, when I was working for De Montfort University, I had my paper accepted for the Digital Earth Summit held in Auckland, New Zealand. It was during this conference that I prepared and submitted my job application for the new Director of

Event	Year	Location	Theme
ISDE 1	1999	Beijing, China	Moving towards Digital Earth
ISDE 2	2001	New Brunswick, Canada	Beyond Information Infrastructure
ISDE 3	2003	Brno, Czech Republic	Information Resources for Global Sustainability
ISDE 4	2005	Tokyo, Japan	Digital Earth as a Global Commons
Digital Earth Summit '06	2006	Auckland, New Zealand	Information Resources for Global Sustainability
ISDE 5	2007	Berkeley & San Francisco, USA	Bringing Digital Earth down to Earth
Digital Earth Summit '08	2008	Potsdam, Germany	Geoinformatics: Tools for Global Change Research
ISDE 6	2009	Beijing, China	Digital Earth in Action
Digital Earth Summit '10	2010	Nessebar, Bulgaria	Future of Digital Earth
ISDE 7	2011	Perth, Western Australia	ISDE7 The Knowledge Generation
Digital Earth Summit '12	2012	Wellington, New Zealand	
ISDE 8	2013	Sarawak, Malaysia	

The International Society for Digital Earth Conference Programme

Richard Simpson
Business Development Director of Nextspace

the Serious Games Institute post at Coventry University. The Auckland conference was another memorable event at which I met more outstanding and influential people, including Richard Simpson, Business Development Director of Nextspace, a leading 3D visualisation company based in Auckland. Richard Simpson has a rich understanding of the potential of 3D visualisation and the concept of Digital Cities. We discussed this and its conceptual connection with work his company had been doing on virtualising the human body as a path to visualising and better understanding the mechanics behind life. New Zealand is almost uniquely placed as a location to study Digital Earth because of its position in the Southern Hemisphere and its environment.

The following year I was invited to speak at and co-chair a workshop at the International Symposium for Digital Earth held at Berkeley University in California. By this time I was the newly-

Doctor Tim Foresman – Conference Director of ISDE5

appointed Director of the Serious Games Institute at Coventry University and had been corresponding with Dr Tim Foresman the Conference's Programme Director since I met him in Brno at ISDE3 in 2003. Tim had organised the most amazing assembly of speakers and themes that I had ever encountered. I was staying in San Francisco which was some miles from the conference venue but I was fortunate to get to know my workshop Co-Chair, Melanie St James, who, as a resident of California, happened to have her beloved yellow VW Beetle car with her and was willing to ferry me to and from the conference. It was at this conference that I got to meet and talk to people who will stay in my memory for a lifetime. One of these people was the Apollo 14 Astronaut Edgar Mitchell, the 8th human being to walk on the moon, who showed a series of slides which included a whole section of photographs he had taken himself both on the surface of the moon and from the module that brought them safely home. He explained that his crew were originally supposed to fly the fated Apollo 13 mission that gripped the world with its drama and incredible achievement in the face of adversity but a slight illness of one of his crew had meant that the reserve crew had to be drafted in. He talked about his life and burning ambition as a child to fly aircraft in the early days of aviation and the most emotive slides for me were the series he titled "Coming Home", taken from the lunar module window as it rotated every 2 minutes to display in turn a receding moon and a growing spectacular earth. There were tears in my eyes as he read a poem he had written in the few weeks after he had returned to earth to articulate the minuteness of mankind compared to the infinity of space.

It was during ISDE5 that I also had the privilege of listening to and having lunch with Douglas Engelbart, the inventor of one the most

Douglas Engelbart – Inventor of the Computer Mouse

important interface technologies of the 20th century, the computer mouse. In his presentation, he showed the original black and white video from 1968 of what became known as the "Mother of all Demonstrations" in which he showed it being used to control a computer and also access hypertext facilities. Doug, even in his 80s, was a sprightly, highly intelligent man who was still full of ideas and passionate about using his abilities to try to make the world a better place. Doug was responsible in his lifetime for two of the most significant developments to transform the ability of machines to interact with us in a more intuitive way. The mouse was a major advance in making communication with a computer more accessible to the human race through a better interface, and hypertext improved the ability of the computer to communicate with human beings. Hypertext which effectively needs to be "programmed" and managed by human beings was really a forerunner of search engine technology which today is self-generating and increasingly intelligent. Machines today have the capacity to learn from the millions of searches we make every second and apply that intelligence to improve the relevance and effectiveness of the search process.

One of the other people I met at ISDE5 went by the name of "Planetwalker". His real name is John Francis and he was a member of the discussion panel on indigenous communities and citizen engagement for sustainable development. As he was entertaining the conference with his songs, Planetwalker told his remarkable story which began in 1971 with the San Francisco bay oil spill disaster when two Standard Oil tankers, the Oregon Standard and the Arizona Standard, collided and 800,000 gallons of oil was spilt into the Bay Area causing massive damage to wildlife and the environment.

John Francis, 'Planetwalker'

John Francis witnessed the after effects of this spill and vowed to play his part in preventing such a tragedy happening again. He began by not using any method of transport which consumed fossil fuels so he would walk or use a bicycle to travel anywhere. His parents lived quite a distance away and so he would have regular phone calls home to his parents. He adjusted to walking everywhere very quickly and wanted to do more to help the environment so he decided to stop talking. His description of his last phone call home to tell his mom that he wouldn't be phoning her in future was hilarious but after he decided to take this drastic step he managed to get his PhD and even became a lecturer at Berkeley University and later on a United Nations Ambassador for Oil and the environment, all without talking. He did not speak for many years before deciding to have a celebration party with his family after which he began to speak again. He claimed to have had great value from this period of his life, not least because he began to listen to other people far more instead of interrupting them or not really hearing them. Despite this bizarre and rather extraordinary story, Planetwalker came over as a very intelligent, humorous and warm human being who, like Thomas Cook, had achieved extraordinary results through a passion for mankind.

The Digital Earth Symposia focus on the use of sensor technologies and sophisticated visual interfaces like Google Earth to help mankind engage in a two way communication with our planet or "Spaceship Earth" as Bucky Fuller called it. The sensor technologies detect both

Speaking at the Indigenous Community panel session at Berkeley in 2007

Rebecca Moore presenting on Google Earth at ISDE5 in 2007

location and parameters which reflect the health of our planet such as ocean temperatures, pollution levels etc and this information is used by the immersive technologies in Chapter Nine to communicate back to human beings in four dimensions (space and time) the consequences of modern life. As Co-Chair of the ISDE5 panel on indigenous communities, I focused on the use of serious games for sustainable development and the importance of engaging communities at a local level in positive actions not only to analyse and act on information, but also to play an active role in becoming "human sensors" willing and able to contribute to this body of knowledge through our own observations.

One of the other presenters at ISDE in Berkeley was Rebecca Moore who was presenting the Google Earth Outreach programme she conceived and leads at Google and which supports nonprofits, communities and indigenous peoples around the world in applying Google's mapping tools to the world's pressing problems in areas such as environmental conservation, human rights and creating a sustainable society. She is a computer scientist and longtime software professional. Her personal work using Google Earth that she discussed at ISDE 5 was instrumental in stopping the logging of more than a thousand acres of redwoods in her Santa Cruz Mountain community. Today, Rebecca also leads the development of the Google Earth Engine, a new technology

Melanie St James – ISDE Co-Chair and Founder of Empowerment Works

platform which supports global-scale data mining of satellite imagery for societal benefit.

The Co-Chair of the ISDE panel on indigenous communities was Melanie St James, the social entrepreneur Founder of a global initiative called "Empowerment Works". Melanie also ran some practical workshops on community engagement and has played a major role in raising awareness globally about the environment and the part technology can play in solving some of these complex issues.

It was during ISDE5 in Berkeley that I was also introduced to Professor Manfred Ehlers who had presented a paper on the integration of physical and virtual worlds that his university had been exploring within their own campus. Manfred had been tasked with organising the next Digital Earth Summit in his native Germany at Potsdam. I attended this summit and presented a paper. The summit took place not long after a major earthquake disaster in China and included an impressive presentation by Professor Guo Huadong, Director General of the Chinese Centre for Earth Observation and Digital Earth, on the use of remote sensing and imaging satellites for disaster management.

In September 2009 the International Symposium on Digital Earth (ISDE6) with the theme of "Digital Earth in Action" was hosted by International Society for Digital Earth and Chinese Academy of Sciences, and organized by the Chinese National Committee of International Society for Digital Earth and Centre for Earth Observation and Digital Earth Science. More than 1000 scholars, enterprises and managerial experts from 40-odd countries attended the conference. Through further research and exchange of Digital Earth theory, technology and applications, the ISDE6 aimed to exert important and positive effects on awareness of Digital Earth's roles in global change and promotion of Digital Earth's service for human survival and development. On September 12th 2009, I was present in the auditorium when the conference announced the Beijing Declaration on Digital Earth which states:

> "We scientists, engineers, educators, entrepreneurs, managers, administrators and representatives of civil societies from more than forty countries, international organizations and NGOs, once again, have assembled here, in the historic city of Beijing, to attend the Sixth International Symposium on Digital Earth, organized by the International Society for Digital Earth and the Chinese Academy of Sciences, with co-sponsorship of sixteen Chinese Government Departments, Institutions and international organizations, being held from September 9-12, 2009:
>
> Noting That
> Significant global-scale developments on Digital Earth science and technology have been made over the past ten years, and parallel advances in space information technology, communication network technology, high-performance computing, and Earth System Science have resulted in the rise of a Digital Earth data-sharing platform for public and commercial purposes, so that now Digital Earth is accessible by hundreds of millions, thus changing both the production and lifestyle of mankind;
>
> Recognizing
> The contributions to Digital Earth made by the host countries of the previous International Symposia on Digital Earth since November 1999, including China, Canada, the Czech Republic, Japan and the USA, and by the host countries of the previous Summit Conferences on Digital Earth, including

New Zealand and Germany, for the success of the meetings as well as further promotion of Digital Earth;

Further, that the establishment of the International Society for Digital Earth and the accomplishments of its Executive Committee, the launch of the International Journal on Digital Earth, and its global contribution to cooperation and data exchange;

That the themes of the previous seven meetings: Moving towards Digital Earth, Beyond Information Infrastructure, Information Resources for Global Sustainability, Digital Earth as Global Commons, Bring Digital Earth down to Earth, Digital Earth and Sustainability, Digital Earth and Global Change, and Digital Earth in Action, have laid out a panoramic scenario for the future growth of Digital Earth;

That Digital Earth will be asked to bear increased responsibilities in the years to come, in the face of the problems of sustainable development;

Further Recognizing
That Digital Earth should play a strategic and sustainable role in addressing such challenges to human society as natural resource depletion, food and water insecurity, energy shortages, environmental degradation, natural disasters response, population explosion, and, in particular, global climate change;

That the purpose and mission of the World Information Summit of 2007, the Global Earth Observation System Conference of 2007, and the upcoming United Nations Climate Change Conference of 2009, and that Digital Earth is committed to continued close cooperation with other scientific disciplines;

Realizing
That Digital Earth is an integral part of other advanced technologies including: earth observation, geo-information systems, global positioning systems, communication networks, sensor webs, electromagnetic identifiers, virtual reality, grid computation, etc. It is seen as a global strategic contributor to scientific and technological developments, and will be a catalyst in finding solutions to international scientific and societal issues;

We Recommend
a) That Digital Earth expand its role in accelerating information transfer from theoretical discussions to applications using the emerging spatial data infrastructures worldwide, in particular, in all fields related to global climate change, natural disaster prevention and response, new energy-source development, agricultural and food security, and urban planning and management;

b) Further, that every effort be undertaken to increase the capacity for information resource-sharing and the transformation of raw data to practical information and applications, and developed and developing countries accelerate their programs to assist less-developed countries to enable them

Professor Guo Huadong announcing the Beijing Declaration on Digital Earth

to close the digital gap and enable information sharing;

c) Also, that in constructing the Digital Earth system, efforts must be made to take full advantage of next-generation technologies, including: earth observation, networking, database searching, navigation, and cloud computing to increase service to the public and decrease costs;

d) Further, that the International Society for Digital Earth periodically take the lead in coordinating global scientific research, consultations and popular science promotion to promote the development of Digital Earth;

e) Expanding cooperation and collaboration between the International Society for Digital Earth and the international community, in particular with inter-governmental organizations, and international non-governmental organizations;

f) Extending cooperation and integration with Government Departments, the international Scientific and Educational community, businesses and companies engaged in the establishment of Digital Earth;

We Call for
Support from planners and decision-makers at all levels in developing plans, policies, regulations, standards and criteria related to Digital Earth, and appropriate investments in scientific research, technology development, education, and popular promotion of the benefits of Digital Earth."

The Beijing conference was a massive event and I was involved both as a speaker and a co-chair of one of the breakout workshop sessions. The conference also gave me the opportunity to make my first visit to the Great Wall of China.

The following year the ISDE Summit was held in the Bulgarian

Temenoujka Bandrova, Richard Simpson, Milan Konecny and Peter Woodgate at the 3rd ISDE Summit in Nessebar Bulgaria

Summer Resort town of Nessebar with the organisation shared by Professors Temenoujka Bandrova and Milan Konecny. Apart from the tremendous cultural activities which included an evening's traditional folk entertainment in a nearby village, one of the highlights of this conference was a presentation by Alessandro Annoni of the Institute for Environment and Sustainability at the European Joint Research Council (JRC) in Ispra, Italy. His paper was entitled "Digital Earth's nervous system and Volunteered Geographic Information sensing: towards a self-aware planet." The abstract of his paper is quoted below:

> *Digital Earth is a powerful metaphor for the organisation and access to digital information through a multi-scale 3D representation of the globe. Progress made since Al Gore's speech gave a concrete body to this vision. However, this body is not yet self-aware: a better integration of the temporal and voluntary dimension is needed to better portray the event-based nature of our world. We thus aim to extend Digital Earth vision to include a digital nervous system of the globe, providing decision makers with alerts on events of both a known and unknown nature. Such timely, event-based, information has practical applications for crisis management. In order to respond effectively to a crisis situation, managers need up-to-date situational awareness,*

which is traditionally built through trusted information sources (e.g. police, civil protection, media reports, etc.). Recent research highlighted the important role citizens can play on disaster sites by providing geo-referenced information, known as Volunteered Geographic Information (VGI) in real time, as a complement to traditional sources. Although workflows have been successfully implemented to create, validate and distribute VGI-based datasets for various thematic domains, the issue of exploiting VGI in real time and its integration into existing concepts of Digital Earth, such as Spatial Data Infrastructures, still needs to be further addressed. In this paper we suggest to bridge this gap, by proposing the development of sensor web enablement for VGI. In this way, VGI sensing becomes a sense of the Digital Earth's Nervous System.

My own paper at this conference was on the topic of the Social Networking and Immersive Applications for Community Engagement that I will cover more comprehensively in Chapter Nine.

The ISDE 7 International Symposium held in Perth, Australia, was a reflection on how far and how quickly developments in sensor and location based technologies have advanced in the 12 years since the very first ISDE Symposium in Beijing. The content of the presentations showed that the technologies we now have available to measure the earth's spatial and environmental data are almost mature enough to be taken for granted and that the emphasis has shifted from measuring data

The ISDE7 Welcome Ceremony in Perth Australia August 2011

to analysing, visually representing, interpreting and disseminating that information in ways that engage, inform and persuade all sectors of society into positive action for sustainable development.

Machine Awareness of the Dimensions of Space, Time and State

Until the 1980s, the only way that machines were able to be location-aware was through human intervention by programming them with a location identity stored in digital format, whether that was in an ATM or a retail point of sale system or in any of the other devices where physical location was important for any interactions between machines and humans. Prior to the 1980s there was no global positioning system to cover the whole globe although the use of radio signals transmitted from a network of ground stations called beacons had been used in the aviation industry to help navigation and air traffic control through a process of triangulation in which the strength of signals from 3 different beacons could be used to calculate a fairly precise position. The decision to create a satellite based global positioning system was not taken until 1973 when the US Air Force and US Navy committed to the establishment of the systems we take for granted today, although at that time it was intended only for military purposes (http://www.kowoma.de/en/gps/history.htm). It wasn't until after 1983 when a Korean Airline Flight accidentally got shot down over Russian airspace when it had got lost that the decision was taken to allow civilian use. It wasn't until the year 2000 that full availability of the satellite network was put into place for civilian users, improving the accuracy from 100 metres to about 20 metres.

Fiat Blue 'n Me satellite navigation system

Model of Nitrogen Dioxide Concentration in Leicester

This ability to determine precise location from space was not a new idea because ancient mariners had used a device called a sextant for centuries before in order to calculate their position from celestial bodies like the sun or the stars. Once the GPS system of satellites in orbits of about 12,000 miles above earth was in place and the benefits were being realised by users on earth, other initiatives from nations outside the USA have begun their own programmes with aims of higher accuracy. This positioning capability of satellites orbiting the earth every 12 hours can also be used to provide high resolution images of our planet.

My first involvement with companies who make use of this combination of sensor and location technologies was whilst I was working as a Project Manager at De Montfort University in 2005-6 when I began to explore the possibility of using location aware imaging combined with 3D modelling to create a virtual model of Leicester City Centre. These embryonic investigations brought me an introduction to a Leicester company called Infoterra who specialise in airborne and space laser and photographic imaging to provide some of the data which Google use in their applications. It was also through these investigations that I became aware of Google Earth and I was one of the very earliest UK users to download Google Earth Pro for academic use in 2005. Today Google Earth has many millions of users globally not only able to access very high quality visualisations but also to contribute their own data and models.

Professor Madeleine Atkins speaking at the World's First Serious Virtual Worlds Conference in 2007

In November of 2006 I started a new job at Coventry University with the task of setting up the Serious Games Institute (SGI) and establishing it as both a global thought leader and a commercially sustainable operation. Contract signatures for the capital expenditure to acquire and equip the SGI building were signed at the end of March 2007 and the project was officially launched at the world's first Serious Virtual Worlds Conference in September 2007. The Immersive Technologies that the SGI focuses on and the story of SGI's development are more fully covered in Chapter Nine but there were over 100 physical attendees in Coventry for the launch conference supplemented by a virtual audience participating in the virtual world called Second Life. By the time of the conference I was convinced of the potential of integrating physical and virtual worlds as a means of addressing many of the serious social, economic and environmental challenges the world faces and that the essential technologies for enabling machines to accurately sense space, time and state were mature enough to build applications around these challenges.

Wireless and wired sensor technologies were able to communicate the dimension of "state" i.e. properties such as temperature, humidity, energy usage etc, GPS and location aware technologies could pinpoint spatial position accurately and synchronised time stamps brought knowledge of temporal position. By connecting these technologies into

Microsoft Kinect intelligent interface device

a 3D virtualisation of the real world, mankind could at last potentially conquer these multi-dimensional barriers. All the components are now pretty much in place to enable machines to assist our understanding and analysis of the physical world and, through the simulation technologies in Chapter Nine, more accurately predict the future. Machines can communicate to us through a variety of interfaces, at an increasingly personal level, what has happened in the past, what is happening now and what is likely to happen in the future.

In order to use this knowledge to create a sustainable future, mankind is equipping itself with a range of sensor and interface technologies that can be used to "control and instruct" machines to carry out our will and every individual citizen with access to these technologies can now potentially communicate to and influence a global audience.

One of the latest and potentially most powerful interface devices is the Microsoft Kinect which was developed primarily as a controller for video games but which has massive capability to transform the human machine interface. The Kinect has three dimensional "eyes" and "ears" that enable it to recognise human faces and human speech and to track and replicate physical motion. This technology, although still embryonic, is one of the last frontiers between the domains of humans and machines and in crossing this frontier, many of the barriers of space, time and state will be broken. We are entering the final technology chapter of my journey and yet, in many ways, our trip has only just begun.

CHAPTER NINE
It's Only a Game

This book's journey has brought us at last to the beginning of a new era in which we are likely to discover whether all the intelligence we have been transferring to machines to enable us to understand and conquer the physical constraints of our universe will liberate us or destroy us.

"Immersive technologies" is a relatively new description for the digital media applications that we are today choosing to satisfy our basic needs for engagement, motivation, challenge, achievement, risk and reward. These basic needs are born of the same human characteristics that drive us in our attempts conquer space, time and state. My phrase for what has always, in many different types of situation, best engaged our attention and shaped our development is "The same but different". We need the paradoxical combination of familiarity and novelty to make us feel comfortable yet interested enough to explore. Too much familiarity brings boredom and contempt yet too much novelty can cause rejection, dismissal or even stress. This formula is one of the reasons why regular British TV soap operas like "Coronation Street" and "Eastenders" are so popular. The writers and Directors of these programmes make the characters and situations familiar and believable yet the storylines bring new and unexpected twists that keep regular viewers glued to their TV sets each episode.

Human beings seem to be attracted by the co-existence of concepts and attributes that are intuitively contradictory e.g. gentle and

Guitar Hero?

giant. This is the reason why "Serious Games", which along with virtual worlds and social networking make up the main components of immersive technologies, is such an appropriate phrase. These applications are what are now the most successful activities for engaging our discretionary time, attention and money and, because of this engagement, they are acting as catalysts in the acceleration of innovation and change to an unprecedented rate. In essence, from a technology perspective, we are getting exactly what we have wished for and as saying goes "Be careful what you wish for – you might just get it". This chapter explores my personal journey through serious games and immersive applications and sets the scene for the final chapter in which I speculate on what might be in store if we follow this path to its final destination.

The idea for the strapline for this chapter stems from a quote from a famous football manager noted for his dry humour and wisdom. Bill Shankly was the Scottish Manager of the English First Division Liverpool Football Club. He established his soccer team as one of the giants of English and European football in the 1970s. He was a man with a real understanding of human nature and someone whose words could inspire and motivate soccer players like few other managers could. His quotation "It's not a matter of life and death, it's more serious than that" is the antithesis of "It's only a game" and in true paradoxical fashion, both phrases are simultaneously true yet contradictory. Football, to the people of cities like Liverpool, Manchester, London, and down to the infinite number of towns and villages who put out soccer teams every week, is a real passion. It engages their hearts and minds and brings both the comfort and familiarity of a weekly routine and the excitement of the unknown. We all know that if our favourite team loses that the world will still keep turning and that things in our daily life will be much the same but we still feel real pain at losing and ecstatic joy at winning.

It is my contention that all games are inherently serious to both players and spectators alike. Football for me has been a primary immersive activity throughout my life, combining serious games with virtual worlds and social networking. All these elements existed for me long before it became practical to apply technology to the concept of an immersive activity. My love and passion for football came to a large extent from my Grandfather "Pop" Langford and my Uncle Bill. Both men were avid supporters of Boston United Football Club and the weekly pilgrimage to the old Shodfriars Lane ground (now renamed "York Street") was always much anticipated. My grandfather was such a staunch supporter that if he could not attend a Saturday match e.g. because he had to do overtime in his job on the railways as a Gang Foreman, he would send the club his entrance money, much to the annoyance of my Grandmother. Football to all of us in the family was a very serious game which formed the heart of an important social network and gave us a virtual world of fantasy and imagination in which to explore our dreams of glory. Immersive activities can act as a gateway to worlds in which time can either stand still or fly by and where we can be transported in an instant in our minds to other places and other times and thus satisfy our innate desire to break down the barriers of space and time.

As a child, I played football in our street every day and often joined in more organised kick-abouts on Bowsers field (about a quarter of a mile away from my home) if and when we had enough players to put a match together. It was my Uncle Bill who would most often organise these competitive kick-abouts on Bowsers field where the grass was long and peppered with "cow pats" as additional hazards and where we used our jumpers as goal posts. I still have some scars on my right hand from playing in goal and getting badly cut by the metallic ends of flying bootlaces when I made a save. When I played in goal, I was emulating my Uncle Len who used to play competitively in the Boston and District League as a goalkeeper for a local team called "Cherry Corner", named after the cafe that used to stand on Boston Market Place. I remember going to watch him play in a floodlit match on Boston United's ground in a cup competition. Floodlit matches were always especially exciting for me as there is something special about the atmosphere of a football pitch bathed in bright light against a black sky. I think Len let in several goals that night and he blamed the floodlights for not being able to properly see the old brown leather ball as it flew past him into the net. Those games at Shodfriars Lane were wholly immersive activities that we were all glad to invest our discretionary time, attention and "gate money" on.

The Boston United team of the 1950s were legendary giant killers in the Football Association Cup. Unlike today when there are preliminary rounds to eliminate the smaller teams before the League Clubs become involved, every team in the FA cup went into the first round draw, meaning that Boston's first match could be against teams like Manchester United. Boston's heroes in those days were players like Don and Geoff Hazledene and Geoff Snade. There was the epic giant killing act when Boston travelled to high flying Derby County and trounced them 6-1 at the old Baseball Ground and then went on to play Tottenham Hotspur at White Hart Lane where they lost 4-0. I was too young to travel to those games but I regularly went in the 1960s to see players like Arthur Hukin and Norman Rigby, a hunch-backed centre half who took no prisoners.

As well as sports of all kinds, books are also, for many people, a highly immersive activity in which they can escape the barriers of time and space and, importantly, uniquely contribute their own experience to create a personalised version of any story. I remember getting a football book as a Xmas present in the 1950s. It was a biography of Tom Finney who was a legendary Bolton Wanderers and England player noted not only for his sublime skill but also his sportsmanship. I used to read this book over and over again and even though it was not a work of fiction, it gave me a platform around which I could use my imagination to share his world and his triumphs and hardships. Books, like the medium of radio, do provide opportunities to contribute something of ourselves in the experience, enabling us to be, in a very real sense, active producers instead of just consumers.

Even in the 1950s, there were brand icons who gave their name to products to help generate sales, though their rewards for these "David Beckham" like endorsements would have been very modest. This brand endorsement also supports the immersive experience in a way because we somehow believe that the use of a product or service by one of our iconic heroes is a gateway to another and better world. My very first proper football boots were "Johnny Hancocks" specials. Johnny Hancocks was a diminutive (5ft 4in) footballer with the heart of a lion. He used to play for Wolverhampton Wanderers in the 1950s. The boots that bore his name were the old fashioned kind of boots that came up high on the ankle and had leather studs which my dad had to hammer in using the shoemakers lathe that was kept in the shed. Before every game I had to clean these boots and give them a liberal coating of "Dubbin" wax to waterproof them. The balls we played with in those days were made of brown leather and had a bladder inside that you inflated before tying the laces on the ball to keep the bladder inside. These footballs also had to be protected with Dubbin wax and even this was not enough to prevent the ball becoming sodden and leaden in the winter rain. Heading one of these balls on a wet winter day was very likely to cause a headache.

As I grew older I would travel further afield to play on "proper" pitches like Garfitts Lane where they had permanent marked out pitches and goalposts. Playing in these ad-hoc matches with anybody who happened to be around was my virtual world and social networking. When I was standing between the goalposts on Garfitts Lane my mind was placing me in Wembley Stadium in front of a capacity crowd cheering my heroic saves and my mates were my social network bound together by this common passion and dreams of glory. I even made up my own social networking group, a team I called "Bushy Rovers" for which I used a "John Bull" toy printing set to make up my own membership cards that I gave out to the mates I wanted in my team. Selecting teams in those days involved nominating two captains who would take it in turn to pick their players until only the worst players were left to be chosen. Life could be so cruel to those left until last.

The first "proper" football match I can ever recall playing in was when I was at Staniland Primary School and I played in a game on a pitch beside Boston Railway Station. I can still recall the excitement at being involved and the disappointment of not playing well. It wasn't until I went to Boston Grammar School that I began to climb the ladder of the local soccer scene. I was about 14 years old when I was asked to play for a Sunday League Team called "Wyberton FC". It was managed by the owner of a local fish and chip shop, Cyril Borrill, and he used to collect myself and several other players in his white Ford Anglia bread van where we would all sit on its freezing metallic floor whilst Cyril drove us to away matches. It was whilst I was playing for this team that I had my very first match on the hallowed turf of Boston United's Ground. All I could think of all week before the match was how proud my dad and grandfather would be to see me there. Unlike now, I was a slim and diminutive player and I actually scored a goal in that match from a free kick just outside the penalty area. There was only a small crowd watching this game but they were stunned into silence when the youngest and freshest face youngster on the pitch expertly curled the free kick over the wall of players into the corner of the net in a fashion that would have graced David Beckham. The memory of the disbelief amongst the players and crowd alike will live with me forever.

For my parents, the main immersive activities that they enjoyed and that were their serious games, social network and virtual world were card games like Newmarket and Whist. Their regular Whist Drives at the local scout hut were the equivalent of my weekly excursions to Boston United (and further afield with my Grandfather when he took advantage of the railway concessionary tickets). As a family, especially at seasonal get-togethers, we would all find ourselves dragged into a game of Newmarket. Probably prompted by "Pop" Langford, who had a bit of a gambling streak in him, we would play for money. Our stakes were only small copper coins like one penny and two penny pieces but there was always the challenge and thrill of reward for success. Even in these family games Pop did not like losing and there were many occasions when he, my Uncle Len, my Dad and myself would be kept playing until the early hours of the morning to give Pop a chance to recoup his losses.

In the early 1950s when Britain was still rebuilding after the Second World War, immersive activities were an escape from the humdrum of daily life which, for most people, was quite hard. Money was tight and rationing was still in place for many items. The baby boomer generation came into a world where their parents had lived through the austerity of war and, like my father, had missed out on the prime of their youth. Many parents had been robbed of a further education because of the war and had been deprived of some of the best years of their lives. This, as in the case of my father, cultivated a determination that their children would have all the opportunities that were denied to them because of the war. My dad wanted me to have all the things he missed out on in his life and so both my parents encouraged me to enjoy life to the full and make the most of every opportunity.

It is probably because of this encouragement that I have always been a risk taker. Through my parents' solid home foundation I have felt able and wanted to explore all the world has to offer. My dreams and aspirations have brought me many rich experiences in many parts of the world and I can look back now to see how much the world has changed yet still remains the same. Though the world is a very different place in the 21st century, mankind's hopes, dreams and motivations are still pretty much the same as they were when I was a small boy playing marbles under a street lamp in a typical post-war council housing estate.

The marbles games I used to play on a patch of dirt under the street lamp nearest to my home were a microcosm of the serious games, virtual worlds and social networks that children seem addicted to today. Like so many children of that era, I had a marbles bag in my toy box, bulging with these brightly coloured glass balls of varying sizes. The size was an indication of their value in the marble economy around which our games were based. On this patch of earth under the street lamp we would fashion a hole and the objective of the game was to get all the marbles into the hole under rules which were similar to the game of pool. Each player took turns to nudge a marble with their finger to try to collide with another marble and knock that marble into the hole. If you succeeded in getting the second marble into the hole, you got "Chinks again" which meant you had another turn until you missed out. The player who managed to knock the last marble into the hole was the "Winner takes all" champion who added all the marbles in the hole to their collection. I played marbles with my circle of friends and I dare say that we all dreamed of having a massive marble collection and the status of local champion of our patch. This desire to build dreams and form relationships around games and competition is still there today but the main difference is that in my youth we were all constrained by the dimensions of space, time and state. We could only play together when all of us could be under that streetlamp at the same time and when it wasn't raining. Today, our immersive technologies bring us the opportunity to engage in the same types of activities but unfettered by location, time or condition. The activities are more or less available on demand anytime and anywhere.

When I was old enough, I joined the "Cubs" in our local scouting troupe, the 8th Boston (Wyberton) Scouts whose headquarters were in a very basic building with a corrugated roof and an old wood burning stove that was used for heating in the winter. The old scout hut was on a plot of land next to the railway crossing on Yarborough Road and our weekly meetings where we all turned up in our khaki uniforms with our neckerchief and woggles for the usual ritual of "dib dib dibbing and dob dob dobbing" and developing skills that could earn us badges for our uniform. Every year there would be an annual "Bob a job" week in which every scout would be tasked with earning as much as possible for the troupe by doing odd jobs for people for a "bob" (one shilling equivalent to 5p today). I used to have my "regulars" who were mostly friends of my parents and the most lucrative of which were the Waldron family on London Road who were pretty much guaranteed to give me a pound for cleaning their silver. The old scout hut was eventually superseded by a brand new brick building that the whole community worked together on. My dad, being a carpenter, did a lot of the construction work and we kids helped out as best we could and I remember being in the building when they were fitting asbestos panels and the whole air was foggy with asbestos dust. I have often wondered whether this experience may yet come back to haunt me with health problems later in life.

Being in the scouts and looking forward to the weekly meetings and occasional camping trips to places like Woodhall Spa, Chapel en le Frith and West Runton (near Cromer in Norfolk), brought the pleasure of anticipation, and this rich combination of familiar routine and new experience. After the first couple of camps at Woodhall Spa which was the regular Easter break camp, everyone knew what to pack, how to set up the tent and what the camp looked like but we all knew that every camp would bring some new adventure and memories to share. These were all things that shaped our development and the constraints of space, time and state were an important factor both in our engagement/ motivation/excitement and the pace of our development. The challenges that the world of the 1950s and 1960s imposed on us were at a much more leisurely pace than those facing all of us today and whilst we are constantly developing technologies to help us manage this accelerated pace of change, those technologies continue to add to the challenges.

It was football however that was my main passion through my teenage years into my early 20s. Modest though it might seem by today's

8th Boston (Wyberton) Scouts setting off for camp – I am far left carrying a kitbag

Boston United Reserves 1967-68. I am in the back row between Geoff Snade and Brian Clifton

standards, my ambition was to play for Boston United whom I had regularly supported since I was a small boy. During my days playing for Wyberton FC in the Boston and District Sunday League, I caught the eye of some local football scouts and joined the Boston United Colts Sunday

League team. At this time, Boston United main football club was close to extinction and had all but dropped out of non-league football. The lowest point I can recall was a massive defeat in the FA cup by local rivals Spalding United. The following season, the club's fortunes began to revive when they were able to join the United Counties League and make their first steps back to the former glory days. The first team goalkeeper in those days was Ken Oxford who had played senior football with Norwich City and Derby County in the 1950s. After that first year in the United Counties League, Ken Oxford became manager of the Boston United Reserve Team and I was signed to play for them in the Lincolnshire League. Although we were in the senior Lincolnshire League, the team was also eligible to play in the local Boston and District Cup competition. It was one of these games that provided my most vivid memories of those times. We were drawn against Wrangle FC, a powerful village team on the main road to Skegness. Our first match against them was away and resulted in a draw so we had a replay at Boston United's ground under floodlights. The score was still a draw at full time and remained drawn after extra time. There were no penalty shoot-outs at the time so we tried to finish the match with a "golden goal" i.e. the first team to score won the match. Eventually the match had to finish because the floodlights had to be switched off after a certain time and the score was still a draw so we had another replay! I had a full season with them before Ken Oxford moved on to become Manager of a new rival football team, Boston FC, and he asked me to sign for their Midland League side, a real step up in the non-league football world. I made my debut on the right wing in the first game of the season at another local rival, Grantham FC.

 Whilst I was fulfilling some of my dreams of football success with both Boston clubs, I was also signed up to play for a Boston and District Sunday League team called Gipsy Bridge. The team was managed by Gilbert Sands who worked in the Currys electrical store in Boston's Narrow Bargate. Gilbert had ambitions to make Gipsy Bridge the best team ever to play in the Boston and District Sunday League so he set about recruiting all the best local league players, including some of my team mates from Boston United and Boston FC. Gipsy Bridge had a fantastic record in the season that I played. We won all 26 of our matches and scored a total of 126 goals in the season, conceding about 5 goals all season. There was a great sense of team spirit and humour in the dressing room at this club. Our home pitch was on a field in the middle of nowhere about 6 miles north of Boston in the direction of Horncastle. Our biggest rivals were a local team called Real Towell whose name was a combination of Real from Real Madrid and Towell after the timber merchants of the same name. The world outside of the UKs shores was starting to become more accessible in the late 1960s, especially in the footballing context with the giants of Italian and Spanish football. This breaking down of the continental barriers was reflected in the new names of some of the old village teams such as "Inter Wyberton" and "Sporting Club Bicker" both of whose names, without disrespect to these strong village sides, were a triumph of hope over expectation.

Football certainly occupied a lot of my time, attention and finances during my formative years. It allowed me to explore fantasy worlds in which I would be a hero of a cup-winning team, and it gave me my strongest social network, brought together each week to a common purpose, and all diverse in our backgrounds and education. In the Gipsy Bridge team, most of the players worked locally and few, if any, had aspirations to go to University or travel the world. Some of the sharpest senses of humour were to be found in the manual workers whose skills shone like a beacon every Sunday morning yet who would never achieve any academic heights. Teams and clubs of all kinds are communities that satisfy our need to belong, to share, compete and develop. The pleasure and anticipation of each week with its build up to the match was fertile ground for imagination, dreams and motivation heightened by the fact that there had to be a specific time and place in which we would gather to collectively work out our fantasies and plans. Today we have succeeded in making serious games, virtual worlds and social networks available on demand through our digital and mobile technologies and I wonder if this immediacy and accessibility inflates our expectations and makes us increasingly demanding of richer and more instant gratification.

My soccer exploits continued at University and I was disappointed that my initial soccer trial on Birmingham University playing fields did not gain me entry into either the First or Second XI teams. The trial match was being organised by the Second Team Captain, Dave Barraclough, and I recall that I only played about 20 minutes as there were so many players being tried in rotation. I started my University football with the Third team whose captain was a no nonsense defender called Al Riley. I think it was the fact that I scored a hat trick against Erdington Wednesday that opened up a chance for me to move up to the Second XI and then get my chance in the University First Team. The very first game I was selected for in the 1st XI was as a substitute in one of our regular mid-week games against other University teams. We were playing against Warwick University and I was brought on in the second half of the game and told to mark Steve Heighway who, even though he was still at Warwick University, was a regular first team player in Liverpool's First Division team. He went on to become one of Liverpool's legends and also gave me one of my most memorable moments in a match the following season.

    I made my first full debut for Birmingham in a night match against a professional club, Warley. Soon after this I made the number 11 shirt my own and became a regular for the rest of my 3 years at Birmingham. Our team was one of the best teams in University football and vied with Loughborough Colleges for top spot. We reached the semi finals of the UAU championships in my first year, just losing away at Swansea University and then we made the UAU finals in both my other years, only to lose narrowly to Loughborough Colleges on both occasions. In my second year at Birmingham, I played against Warwick University again in a home match on the University playing fields at Wast Hills. I had a particularly good game that afternoon and the legendary Steve Heighway came into our dressing room after the match to congratulate me on my performance. I related this story to my parents the next time I was home and somehow the story got lost in translation and became interpreted that I was a personal friend of Steve Heighway who would no doubt visit my humble Wyberton home in his Rolls Royce on some future occasion. My parents became Liverpool FC and Steve Heighway fans and the kids in our street kept asking me when he would be visiting me. I was so amused by this turn of events that I wrote to Steve at Liverpool FC and he sent a lovely letter to my parents along with several copies of his autograph for the kids.

I was appointed treasurer of the University Football Club and it was this role that played a part in my learning to speak German. The Club arranged a football tour of Germany with several matches arranged against local teams in the beautiful Heidelberg area. There were about 15 of us in a minibus driven by our first team goalkeeper Ian "Ted" Finch and in our efforts to economise, there were several players sleeping in each room in what became rather smelly and unpleasant conditions. There was excessive consumption of alcohol that led to at least one player being substituted after vomiting at the side of the pitch. The local communities were extremely hospitable to us at the town hall receptions laid on for our benefit. This was despite one of our players, a medical student called Bob "Wasser" Stockley, doing a Basil Fawlty impersonation from the British Comedy Fawlty Towers episode in which Basil insults his German guests by saying "Ze war is all over now, we are all friends".

Sports performance and indeed performance in many immersive activities is, in my experience, hugely influenced by psychology. Enjoyment, confidence and determination all play a big role in our achievement and development. Sports players talk about being in "the zone" – a state in which mind and body are totally focused and in harmony with an activity. Being in the zone represents the ultimate state of total immersion in which the body's senses are almost shut down as part of a process to bring all our resources to a single common purpose, blotting out pain, noise and all unnecessary distractions. In this state, everything seems to "flow" and we become capable of extraordinary and, in some cases, superhuman achievements. I can remember football matches in which everything seemed to go right and I felt able to place the ball anywhere I chose. In my cross country running, perhaps aided by the monotony of the pounding feet, I was able to run mile after mile like a machine without tiring and without thinking about what I was doing.

When I was 19, I ran in the annual Kennedy Memorial Test, a 50 mile marathon from Louth to Skegness and back across the Lincolnshire Wolds. My motivation for taking part in this event was money. A local man had voiced his opinion that no-one in our village could come in the first 20 finishers in this race and he was quoted in the Lincolnshire

Standard as offering £20 to any individual plus £20 to charity for anyone from Wyberton who could finish in the top 10, and half these amounts to anyone who finished between tenth and twentieth. My Uncle Bill had recently been diagnosed with cancer and I had been standing beside him at Boston United Football ground the day he first collapsed and was subsequently mis-diagnosed as being anaemic. His cancer seemed to have stemmed from the shrapnel wounds he got when his Royal Navy ship "The Ivanhoe" was sunk in the war. I wanted to race to win some money for Cancer Research to help my uncle and so entered the race, along with our milkman who was convinced that he would do well because of all the miles he walked every day on his milk round. My milkman gave me a lift to Louth and we set off along with around 560 other runners on a cold April day in 1970. Unbeknown to me, one of the other runners was a 29 year old ex-Olympic sprinter and future celebrity author, Jeffrey Archer.

I started the race at much the same pace as I used to run the school cross country races and found myself in the lead for about the first eleven miles. However, unlike Boston and the surrounding countryside that made up the Lincolnshire district known as South Holland, the route took us up hill and down dale for a large proportion of the 25 miles to Skegness. During this first part of the race, I was pretty much in "the zone", and was able to drive myself on without too much pain until the welcome sight of Skegness Clock Tower came into sight and I realised that I was only half way through the ordeal and I had already been running for over 3 hours. I did manage to continue to jog along to around the mid-way point on the return leg when I began to find myself being overtaken by other runners until I was joined by an older man with whom I struck up conversation and explained why I wanted to finish in the top 10. He ran with me from that point and paced me until he was sure I would achieve my goal and, about 100 yards from the finish line, he accelerated away to finish just ahead of me. I actually finished 7th out of 560 runners in a time of 8 hours 41 minutes and was very satisfied to beat Jeffrey Archer by almost 5 hours but my milkman had to drop out when he reached Skegness. The next day my legs were so cramped I could not walk at all for 3 days. Now, when I go to the gym each morning that I am in the UK, I can just manage to jog on the running machine for

15 minutes at an average speed which is less the average speed I ran the fifty miles of the Kennedy Memorial Test in.

Once I left University I joined the nearest non-league professional soccer club I could find, Loughborough United, who played in the Midland League. The club was managed by Ritchie Barker who had his previous career as a promising professional footballer cut short by injury. Ritchie later went on to far better things to become Manager of Sheffield Wednesday in the days when Chris Waddle played for them. I was made club captain in my first year and finished up the second highest goal scorer in the season. I had some very good times at the club but there were severe financial problems which led to my driving the club minibus to away matches and the club playing two away games in one day to save money. In the second of these games we lost 13-0 and I believe that was the last match the club played before it went into liquidation and the ground was sold to make way for a sports centre.

Once Loughborough United was wound up I signed for Hinckley Athletic in the West Midlands League and it was an injury in the first game of the season that scotched my chances of playing for a national team. I had been selected for the national trials of the Post Office soccer team but I turned over on my ankle in a match the week before the trials and had to drop out. I did captain the Post Office Midland Region Soccer Team and we actually beat the National side in a game played at Stone in Staffordshire. I had a couple of good seasons with Hinckley and was named club player of the year at the annual dinner but it was an injury that I sustained in an away game at somewhere like Hednesford that finally persuaded me to stop playing football. It was a bizarre injury in many ways because the tackle that bent my knee back against its joint did not hurt but when I stood up, I had lost all control over the lower half of my leg, which was swinging like a manic pendulum in all directions. I did have physiotherapy and was able to walk and run soon afterwards but much of the enjoyment had gone so I took up squash instead.

Squash took over as my main immersive activity, serious game, virtual world and social network. I began playing at Wanlip Squash Club just off the A46 north of Leicester and, along with Marion Hughes and Diann Lester, started up a Post Office Squash Club. After a couple of seasons at Wanlip, a brand new club opened up next to Leicester

Racecourse and my squash playing career with Squash Leicester began. This club was the birthplace of my Rentacrowd social network and my introduction to the "Space Invaders" video game. I was a fairly good local squash player and became able to give our County squash players a decent match. We had the World Number Three ranked squash player as club coach, and the club owners, Ian Turley and Richard Wilson, had both been good County players. I became a regular in the Squash Leicester First Team and acquitted myself pretty well in club matches. My biggest claim to fame was that I used to regularly beat Simon Taylor (when he was 14) who went on to become the England Number One Squash champion.

Immersive technologies in my youth were pretty much the domain of what is now known as "Out of Home Entertainment". It was certainly true that radio and television were embryonic forms of immersive technologies that relied on captivating story lines and believable scenarios that listeners and viewers could identify with enough to form their own engaging mental images and emotions. For me, it was the Saturday morning Mickey Mouse Club at the Odeon cinema in Boston and the adventures of Brick Bradford that kept me absorbed from week to week. In the summer time, the Boston May Fair and the annual visits to the seaside at Skegness or Hunstanton with their penny arcades, Ghost train, Caterpillar, Big Wheel, Helter Skelter and Cake Walk rides were what I looked forward to as escapes to new experiences and the excitement of the unknown. All of these activities were fantasy worlds with strong social networking elements that were not available to me on demand on a daily basis and that made them a special source of anticipation.

It wasn't really until the 1970s that arcade games began to become mainstream and ubiquitous. I can remember the early games such as "Pong" that was launched by Atari in 1972 and was one of the very first video games to achieve a substantial player base. My own favourite was the "Space Invaders" arcade game originally launched by Taito of Japan in 1978. It was the only arcade game at my squash club and I confess that I spent many hours in the enjoyable but fruitless pursuit of stopping the rows of little green pixelated alien creatures from descending down the screen to the sound of slowly intensifying thumping noises. These arcade games were not meant to be serious or have any pretence at realistic simulation but they were extremely popular and generated huge sums of cash for their developers and the console manufacturers that ran the software. The financial success of these video games has played a large part in the evolution and development of the most important elements of our technological progress in the last 40 years. Video graphics, 3D rendering, colour displays, processing power and broadband would not have advanced so rapidly had it not been for the influence and popularity of video games. It is the consumer's unceasing demand for ever more engaging applications which deliver the enjoyment, anticipation, escapism and social connections that immersive activities have provided human beings since homo sapiens first walked this earth which drives these improvements in speed, capacity and multi-sensory richness.

I have remained a consumer of immersive experiences all my life but, like the majority of people, I enjoy and need to be able to create immersive experiences for other people to share. I believe that the ability to contribute to engaging other people is a vital part of managing our lives. Engaging and absorbing other human beings is essential to managing and shaping our lives. The more effective we are at capturing the time and attention of the most important people in our lives, the more successful and influential we become. We all need to feel part of activities that can make a difference to our own and the lives of other people. We find it enjoyable and satisfying to be part of a crowd or community with a common purpose and a shared vision. This sense of belonging to a tribe with shared passions is in itself a contributor to the experience of immersion. Watching a football match or a live concert, our presence in the audience, along with everyone else, plays a part in group dynamics, and has done for thousands of years.

The technologies that enable us all to become producers of immersive experiences as well as consumers are really very new. Web 2.0 is the term that was coined for the tools that enable the man or woman in the street to create content that can be accessed by a global audience. Even before Web 2.0, tools were emerging that were designed to support the process of engagement and motivation. In my own working life, it wasn't until after I started my business, Mass Mitec, in 1984 that the graphics design tools that shaped my business became available on the desktop IBM PC. These tools and their development are described in Chapter Five of this book. Early products like Execuvision in 1985 allowed me to design visual content which included simple cartoon graphics that I used to engage attention in my presentations. As well as the power of the simple imagery to communicate, it was the novelty of the medium that helped me get my message across and influence people. Prior to these technologies being available, I had to rely on painting pictures with my words and performance. Whether it was managing my staff in the External Works team, training new supervisors at the Post Office Telecomms Management College in Bexhill or selling IBM computers, I had to almost entirely rely on my personal skills, knowledge, experience and ability to communicate effectively.

The tools to create immersive experiences started to become accessible to the early adopting consumer in the 1980s with the development of home computers like the Sinclair ZX-81and Spectrum computers that kick-started many a back bedroom hobbyist game developer. Mainstream business tools that improved the ability to communicate visually began to appear in the early 1990s with products like Harvard Graphics, Lotus Freelance and Microsoft Powerpoint and with these developments and the launch of the internet, greater numbers of ordinary people began to be empowered with the tools to express themselves to a global audience. At Mass Mitec, I first witnessed to power of serious games for education and engagement at an international Congress on Epilepsy in Prague in the late 1990s. This custom developed game combined multiple choice questions with a memory card game as a tool to attract visitors to a pharmaceutical exhibitor's stand. The success of this game in attracting visitors to the stand led me to engage a student from De Montfort University to develop a generic version of the game that could easily be modified to any client wanting to attract visitors to their stand at an exhibition.

The first "off-the-shelf" software package that we experimented with at Mass Mitec was developed by Corel of Canada and was called "Click and Create". We bought this in around 1997 and it was one of the earliest video games packages capable of being used by non-programmers. The origins of this software were from a French company called "Clickteam". Mass Mitec was commissioned to develop a fun game for the Foundation for Water Research to use on their exhibition stands and our "Water Game" was in a similar style to space invaders. Although Mass Mitec's core business was based around interactive multimedia, our foray into serious video games was not a success, and so it wasn't until I joined De Montfort University (DMU) in the September of 2005 as the Creative Industries Knowledge Network Project Manager that I became involved once more in immersive technologies and the companies that based their business around them.

My job at DMU was to build bridges between the Creative Industries and Higher Education in order to benefit both sectors through transfer of knowledge and experience. DMU themselves had a Virtual Reality "Cave" installation run by Howell Istance in the Faculty of Computer Science and Engineering. When I visited Howell for a demonstration, I had to wear 3D glasses to watch a helicopter flight around a virtual model of new planned developments in Leicester City Centre. I was sat on a chair in front of the large curved screen as the simulation showed the flight from the helicopter perspective and I found that the experience made me feel slightly sick.

DMU also had an excellent Game Art Design facility run by Michael Powell who had good connections into major games industry partners such as Codemasters. I witnessed a lot of creative talent at DMU during my one year with them both amongst the students and some of the Leicester and UK businesses I involved in my programme of seminars, workshops and masterclasses, all of which were designed to bring students, academics and creative businesses together. At this time, Leicester was at the beginning of a regeneration programme designed to revitalise the heart of the city through a massive modern retail complex at High Cross, effectively an extension of the successful Shires complex, and the development of the St George's area of the city as a new cultural quarter with a mixture of creative arts and inner city living.

As part of the cultural quarter development there was the establishment of a creative cluster of businesses in an old Bus Depot on Rutland Street. This new hub for the creative industries was called "The Depot" and offered a mixture of small offices, hot desks, exhibition space and a drop in café. The Depot was managed by Peter Chandler and was, in my opinion, a very successful initiative that not only kick started many new creative businesses but also stimulated a lot of innovation, collaboration and creativity. Over the 12 months that I was working for DMU, I was exposed to a number of small businesses who were pushing back the boundaries of immersive technologies. Many of these businesses I felt had the potential to grow into much larger and profitable companies but one of the challenges that I have observed in my time working with these businesses is that many of the owners and founders of these highly creative and innovative SMEs see their company as a lifestyle business which is more of a hobby and a passion than a source of income.

Amongst the many companies working with immersive technologies during this period was Maelstrom, a small company based not far from DMU in Upper King Street, Leicester. Maelstrom specialises in immersive virtual reality software and has an impressive corporate client base. Their applications range from visualisation of new planned construction projects, through immersive experiences for marketing to the training of parachutists in a virtual simulator. Leicester has a tremendous record of innovation in virtual reality and a company called "Virtuality Ltd" based in Leicester was one of the very first commercial organisations to develop virtual reality machines for both entertainment and education.

One of the companies based at The Depot was called "Low Brow Trash" and they were a very innovative SME working with mixed reality for entertainment and visitor attraction. One of the installations they developed used a kind of "Punch and Judy" booth in which you placed your head, pulled a face, and then had your photograph taken digitally for transfer onto a fantasy creature in an interactive booth in another room. The interactive booth used ceiling mounted cameras to track your movement and control various creatures on a projected display. These creatures had your digital photo used as their face. The technology behind this innovative installation was only embryonic but the concept of personalising an interactive application was quite visionary and a forerunner of new devices like the Micrsoft Kinect.

One of the most innovative companies working with Mobile Technologies for immersive applications was Nottingham based SME Active Ingredient. One of their specialities was location based games using mobile phones and/or Personal Digital Assistants (PDAs). They developed a game called "'Ere be Dragons" to encourage exercise and city exploration. The phrase that gave the game its name came from ancient map making when they used that expression to indicate unknown or unmapped territories. The game ran on a HP PDA linked to a heart monitor to measure your pulse. The game involved tracking a player around the city of Nottingham using GPS tracking technology on the PDA to show relative locations and the display was clearest when the heart monitor showed your pulse at the optimum rate for exercise. If you ran too hard or walked too slowly it made playing the game harder. Another game this company planned to develop was called "Love City" and involved the use of mobile phones and SMS messaging in the Midlands cities of Leicester, Nottingham and Derby. These 3 cities were to be virtually combined into one big city displayed on large screens at the Creative Hubs of each of the cities with players using GPS to identify partners to send messages to in this virtual Love City.

I spent a year at De Montfort University working with both academics and small businesses, organising regular workshops and an international conference at the Marriott Hotel in Leicester with a focus on the role and value of creativity and innovation in regional development. The audio visual support and recording of the event was handled by a creative Leicester enterprise called Channel 2020 who, led by their MD and Founder Rob Potter, had been a pioneer of the digital revolution, especially as a support to establishing Leicester as a global leader in this sector.

It was whilst I was at de Montfort University that I developed links with Oman and a UK citizen called David Pender who, having lived in Oman for about 20 years, was playing an important role in trying to develop Oman's "brand" in a global economy. Oman, home of the legendary Sinbad the Sailor, has made rapid progress both economically and socially in the last 30 years as they have transformed their country to enable it to compete in the 21st century global market place. The oil which has been a bedrock of their society and economy is a finite resource and the rulers of Oman recognised that the country needs to develop modern, innovative and sustainable new businesses capable of building the nation's wealth and providing employment for its young people. In the discussions with Oman, games and immersive technologies were identified as an area with substantial growth potential globally and a sector that would be attractive to the nation's youth as well as providing tools to support their education. Oman, unlike the UK and many of the developed nations where an ageing population is a growing problem, has a demographic population which includes a high proportion of young people and the main concern expressed by the Omani Leaders was that the traditional jobs in Government would no longer be available once oil starts to dry up. So it was that over the following years I continued to collaborate with Oman in helping them to develop this strategy.

My time at De Montfort University was a happy one but my job was on a fixed-term contract and, although I was not looking to move on, I was attracted by a job advertisement in the Times that my boss brought to my attention, thinking that I had already seen it. The job, advertised in the Summer of 2006, was for a new Director post at the Serious Games Institute which was to be set up at Coventry University. Having already run one serious games workshop for DMU and had substantial experience working with games technologies I thought the job would be a perfect match for me and allow me to exercise my passion for enterprise, innovation and regional development. The salary quoted was around double what I was earning at DMU and, without a great deal of confidence that I would be successful, I decided to apply. The deadline for job applications happened to coincide with my participation in a Digital Earth Conference in Auckland, New Zealand and, having left the job application until the last minute, it was almost on a whim that I decided to send my application form by email from my hotel room in Auckland, more in hope than expectation.

When I got back to the UK, I had almost forgotten about the application when I received confirmation that I had been shortlisted for an interview at Coventry University. Knowing the reputation of Coventry's ring road as a sea of lost vehicles, I took the precaution of driving over to the Coventry University Technology Park the weekend before the interview to make sure I knew the best route. I am generally very good at finding my way through cities but I had been burnt once before by Derby Ring Road when I had a previous interview that I almost missed because of traffic and navigation problems.

I was told that I was to attend the Technology Park for the interview along with 4 or 5 other candidates whom I would meet on the day and the interview consisted of a formal presentation that I had to prepare to explain why I would be a good candidate followed by an interview with senior Coventry University staff. When I arrived at the interview, I took note of the other candidate names and background and used this information to "Google" them on my phone whilst I was waiting. Of the candidates who were there for the morning session, I felt that my background might give me a reasonable chance compared to the others but there was one candidate who was due to be interviewed in the afternoon and, as I read his profile from Google search, I was dismayed to see that the experience displayed against his name was much stronger than mine. What I didn't learn till later was that the person I had "googled" had the same name as the other candidate but was not the person to be interviewed.

In my presentation, I focused on my international experience and contact network and my passion for community informatics to show how I would set about establishing a global presence for the new Serious Games Institute (SGI) which I learnt had still to have its funding approved by Advantage West Midlands and although an empty building had been earmarked on the Technology Park, the SGI would almost have to be built from scratch both physically and reputationally. I left Coventry University

The Serious Games Institute Building in 2006

in the afternoon feeling at least that I had done myself reasonable justice in my interview. I had just left the M69 on the way home when my mobile phone rang and I was offered the job subject to references. By the time I got home only 15 minutes later, the references had been accepted and an exciting new chapter in my career was about to start.

I began this latest chapter in my career in November of 2006 by attending a Serious Games Conference in Washington where I met Jude Ower and Dan Licari who were both representing the UK interest in serious games. Back in the UK, I began my induction into Coventry University and to learn about what was expected of my new position. The original idea for the concept of a Serious Games Institute came from Mary Matthews, now a Director at Blitz Games, but in 2005 she was working for Advantage West Midlands. It was Mary who recognised the strength of the games industry in the West Midlands and saw the growing potential of the application of video games to serious issues such as education and training, awareness raising, behaviour changing, simulation and marketing. The West Midlands is home to valuable and leading games developers such as Blitz Games and Codemasters and pioneers of games-based learning like PixeLearning.

My role was to oversee the securing of the contract, design the layout and equipping of the building, build commercial partnerships with key stakeholders, recruit staff , establish a global presence for the SGI and develop its commercial sustainability. From the beginning I had some great help from people like Henry Jerwood and Mike Eliot-Higgit, the construction Project Manager. There were a number of important changes that I wanted to make to the original plans and strategy for the SGI's development. First of all, I felt it was important to include other closely-related technologies within the remit of the SGI. These included virtual worlds and social networks which I felt were collectively the growing focus of people's discretionary time, attention and money. I decided that although the Institute was appropriately named, it should encompass a wider range of associated technologies which I branded "Immersive Technologies"

The second influence I sought to exert on the shape of the SGIs development was the layout and equipping of the building. The original plans had been drawn up to locate small businesses and a refreshment area on the ground floor and I felt very strongly that for the SGI to be effective in establishing itself as a hub and international centre of excellence, the ground floor should be primarily used for demonstration and showcasing state of the art technology and applications, especially those developed by local businesses and the university researchers. I set aside the first floor to provide business incubation and office space and to act as a "creative cluster" and laid out the top floor as an Applied Research centre. I got my ideas for an open plan layout on the top floor from a visit to the Knowledge Lab in London where I also met Dr Sara de Freitas for the first time at a workshop event on virtual worlds.

During my first year in post I also set about establishing the SGI as an international thought leader through international conference presentations and PR and publicity organised by the University's PR Agency who made excellent use of some of the innovations that we incorporated into the SGI. The telecommunications and digital media infrastructure also made a valuable contribution to creating the right kind of environment to attract publicity, investment and tenants. The contract for the SGI was only signed at the end of March 2007 and the launch of the Serious Games Institute took place only six months later at the world's first serious virtual worlds conference, organised in partnership with Ambient Performance and supported by some great work by West Midlands SMEs Daden's MD David Burden, Alex Jevremovic and Paul Turner of the WalkinWeb.

My avatar 'Hobson Hagard; outside the SGI building in Second Life

The conference was a big success with over 100 delegates from all over the UK and abroad, and featuring live streaming of the conference in the virtual world Second Life with the virtual Coventry University campus officially opened by the Vice Chancellor Madeleine Atkins. At the time of the launch, all the new SMEs offices were allocated to small companies and I had helped to recruit Dr Sara de Freitas as the new Applied Research Director to work alongside me. I had been impressed by Sara's ambition and reputation and felt that she would do a very good job in building bridges with the academics at Coventry University.

The years that followed saw the reputation of the SGI established regionally, nationally and internationally through the hard work of all those involved in helping to develop the concept. There were some extremely talented individuals to work with and the SGI's development was greatly assisted by the recruitment of Jacqueline Cawston as Senior Business Development Manager and the involvement of Tim Luft in a role as Commercial Director. In retrospect, some things could have been done differently but given the circumstances of the limitations of an existing building and a very embryonic market, I am very proud of what has been achieved in the last 4 years.

Immersive Technologies for Training and Education

The notions of immersion and virtuality have always been important components that contribute to the effectiveness of training activities. Immersion is closely related to focus, concentration, motivation, engagement, absorption and enjoyment, all of which are relevant to achieving learning goals. Before any kind of educational technology, including written communication, existed, mankind had passed on skills and knowledge from generation to generation through the engaging stories told by tribal elders. Good story-telling, narrative and an ability to transport learners to "another space" have been the hallmark of good teaching practices for centuries and today, even with the advanced technological tools at our disposal, these skills are still necessary in any training activity.

The birth of the internet in 1992 and the subsequent development

Serious Virtual Worlds Conference in 2010 held in Second Life Virtual World

and convergence of many digital media technologies including broadband, image processing, wireless, mobile, video games and social networks has led to what has been broadly described as "Immersive Technologies". This phrase is very appropriate because it acts as a category that can encompass the portfolio of technologies to which we choose to dedicate a significant portion of our time, attention and money. We choose to do this because these technologies engage us and deliver benefits which justify this investment.

The principal application areas that have these engaging characteristics are video games, virtual worlds and social networks and it is because they are so successful in capturing our attention, influencing our behaviour and developing our skills that many different stakeholders are beginning to explore how to leverage this phenomenon to meet their own objectives. These stakeholders include commercial organisations using gamification to generate customer attraction and retention, political parties seeking to influence public opinion, NGOs trying to influence citizen behaviour to tackle the most serious issues facing society and educators and training professionals wanting to develop more effective ways of skill development and knowledge acquisition and retention.

Like all new technologies seized on in their early development there is an inevitable amount of hype and unrealised expectations amongst the early adopters and this, I believe, is because many people lose sight of the outcomes they wish to achieve and how (or even whether) these new practices can deliver any improvements. In this section I seek to briefly examine these three primary categories of application that comprise what we understand as immersive technologies.

Video Games / Serious Games

The first video game to become acknowledged as the origin of the apparent oxymoron "serious game" was America's Army which was originally commissioned by the US military as an entertainment game designed to attract teenagers into their recruitment programme. This initial use of the game as a marketing tool led to a realisation that these same technologies might also be applied to training soldiers for combat situations. This realisation led to the genre of serious game development in which the traditional ways of developing skills are either impractical, too dangerous or too costly. Military, Emergency and Medical sectors are the most obvious examples of the sectors where this would obviously apply.

America's Army and most other serious games developed so far have been custom designed by specialist companies with the necessary portfolio of multi-disciplinary skills but now we are seeing the launch of serious game development toolkits like Caspian Learning's Thinking Worlds platform that makes it possible for training professionals to develop their own solutions.

America's Army Serious Game

Caspian Learning Thinking Worlds Platform used for Miltary Training

Virtual Worlds

Using very similar technologies to video games, virtual worlds provide an immersive virtual environment that can be a mirror image of a real world or a completely fantasy world which takes its characters out of familiar spaces into an environment that could stimulate "out-of-the-box" thinking. Virtual worlds have been used for military training, especially in situations where soldiers need to be familiarised with a foreign territory, its people and their customs before active duty.

Roma Nova virtual world simulation of Rome in 400 AD

Virtual worlds are not constrained by the rules and restrictions that usually apply to true games technologies. Characters can move freely to explore these worlds and as well as training applications that provide trainees with orientation tools, virtual worlds are also in active use as a more immersive version of the virtual classroom technologies like Cisco's Webex. In my opinion, the tools that are currently available in virtual worlds for managing virtual classroom content and interaction are inferior to what is used in tools like Webex and Live Meeting.

Social Networks

At first glance, social networks might seem to be the odd man out as their primary purpose seems to be sharing pictures and activities with your friends. However, an increasing number of corporate organisations like BT, Intel and Dassault Systemes have developed and implemented their own custom in-house social networks as a knowledge management tool which can connect subject matter experts at any level in the organisation to those who need their skills and knowledge.

Today, we live in a fast-moving environment in which immersive technologies and serious games have been largely accepted as key

Intel Planet Blue social network

influences in shaping both business and society. The next few months and years will see some very significant opportunities and challenges as these disruptive technologies begin to exert more influence. The world has already been changed beyond recognition since my early life in the Lincolnshire Town of Boston and I would never have dreamt in my teenage years that I, like so many others today, would meet my partner virtually over the internet through a Web Dating Site or that I would see regular changes of career as a natural part of life today.

There is much to be thankful for to have lived through these times and to see the changing role and value of games and immersive activities in daily life from being pure entertainment and escapism to becoming a force that can shape global business and society. "Gamification" is here to stay !!

CHAPTER TEN

End of the Beginning or Beginning of the End?

"It was the best of times, it was the worst of times, it was the age of wisdom, it was the age of foolishness, it was the epoch of belief, it was the epoch of incredulity, it was the season of Light, it was the season of Darkness, it was the spring of hope, it was the winter of despair, we had everything before us, we had nothing before us, we were all going direct to heaven, we were all going direct the other way – in short, the period was so far like the present period, that some of its noisiest authorities insisted on its being received, for good or for evil, in the superlative degree of comparison only."
Charles Dickens – Tale of Two Cities

End of the Beginning

The last fifteen years up to the writing of this book have seen absolutely the most disruptive technology developments in mankind's history and the immediate future is likely to see even more disruptive changes across all aspects of our global society in economy, politics, education, business and inter-personal relationships of all kinds. This is the first epoch to have empowered individual citizens with not only the infrastructure to communicate to a global audience but also the tools to create persuasive content and build personal communities of interest, the so-called "Prosumer" age. Likely developments into the future will not be so much about increasing the speed, reach and capacity of our networks and processors as about the dramatic shift in our relationship with technology, our dependence upon it and the influence technology will have on our humanity and our capabilities.

Speed, Power and Capacity

Moore's Law on the price performance of computing technologies is likely to continue to apply for the foreseeable future as devices get faster, smaller and have greater memory and storage capacity. Our telecommunications networks with their fibre optics already have substantial spare capacity and wireless communications will see major improvements in speed to match broadband standards. Sensor technologies are becoming ubiquitous and cover the full spectrum of human senses as we move towards "the internet of things". We will continue to be attracted by devices that deliver more powerful and richer experiences and manufacturers will continue to be innovative about developing new products and services that "force" the consumer to continuously upgrade and/or replace technology within ever shortening timescales. These trends are the logical extension of how the different technologies described in this book have evolved over the last thirty years but it is my opinion that the most significant developments are likely to revolve around personalisation and convergence as technologies from one application area move across boundaries to be integrated into other areas. We have seen this trend with mobile phones having digital cameras, music players and video games integrated into the latest devices, and cars having Bluetooth voice control and entertainment.

Following the structure of the book, this chapter will speculate on how these trends of personalisation and convergence are likely to impact each application area.

Transport Technology

Transport is a major problem for environmental sustainability. New forms of propulsion that are more efficient and do not rely on fossil fuels will continue to be a major focus in all forms of transport. We have built a society in which, in the developed world at least, there is almost ubiquitous access to personal and public transport technology. We have become accustomed to affordable international travel, low cost imported products from distant lands and personal vehicles and even though all these facilities are a serious threat to the environment, it would be politically almost impossible to make any drastic policy changes that would restrict our access or impose even higher tariffs or disincentives.

All the industries involved in the manufacture or provision of services for transportation will continue to find innovative ways to reduce the environmental impact of transport and the communications industries will in turn focus on richer forms of virtual meetings for business and social encounters that reduce the need for travel. This pincer movement may not make a substantial difference to the level of transport activity because the immersive technologies being developed will paradoxically fuel the demand for the physical movement of people and goods.

All forms of transport, both private and public, will make increasing use of personalisation, convergence and artificial intelligence to make the travel experience richer. Chapter One has already given some pointers on how these developments will manifest themselves in cars and other personal transport devices. The technologies in my own modest Fiat 500 are the tip of a very big iceberg that will not only make personal transportation easier and more enjoyable, but will improve safety and begin to provide some of the controls for society that enable the tracking of transportation usage by individual citizens. This tracking will be an important part of reconnecting citizen rights and responsibilities and building better links between usage and costs. Personal transport technology like cars will be able to adjust itself to the preferences and abilities of the driver.

Public transport systems and infrastructure will embed increasing amounts of sensor technology for automatic recognition of individual citizens as a way of not only automating many billing functions and eliminating much of the costs and burdens of ticketing but also of beginning to deliver personalised services on public as well as private transport. Intelligent transport systems and infrastructure will manage the flow of traffic dynamically and provide transport users with directions for their journeys and automatic reservation of parking space. Cars, buses and trains will have embedded intelligent technology to improve safety through speed management and collision avoidance.

Many of the developments described above are already either technically possible or under development with the aim that every citizen will remain continuously connected to all their social and business networks 24 hours a day, 365 days a year.

Communications Technology

The aim of the next generation of these technologies will be to provide ubiquitous and on demand access to a full and rich portfolio of communication services that are personalised to the individual. Mobile devices are likely to become the gateway that connects our physical and virtual worlds. Phones already integrate telephony, video, music, imaging, video games, 3D visualisation, augmented reality, embedded sensors, GPS and computing technologies in a single handheld device and tablets, with their larger screens, are already competing to provide the same levels of functionality but with larger screens which are more suitable for applications such as movies, electronic books and video conferencing.

Some manufacturers such as Blackberry are implementing encryption technologies into their mobile phones to protect the security of personal communications but it is my view that for society to have a safe and sustainable future it will be essential to develop ways to uniquely and securely connect citizen digital identity to physical identity. Whilst this is already seen as essential to secure electronic commerce, it is a politically sensitive area for all other forms of personal communication where privacy and anonymity are viewed as essential human rights.

Personalisation of communication devices to the individual is a natural extension to current trends. Our mobile devices will become an electronic alter-ego which contains and maintains our digital lives. These devices are likely to develop technologies which uniquely connect them to their owner/user in ways which go beyond password access to the device. More advanced forms of biometrics could be developed so that our mobile devices can only be used by the owner, not only protecting that owner from theft but also ensuring that the origins of any communications can be verified. I envisage that our mobile devices will develop sophisticated personal profiling systems similar to those evolving in social networking software so that we can more effectively control our communications and our relationships not just with other human beings but also physical spaces and objects.

Computing Technologies

During the span of my lifetime, the architecture of computing resources has oscillated back and forth between centralised processing/storage and desktop processing/storage and hybrid combinations of both. Right up until the 1980s, mainframe computers were king until the desktop personal computer began to place computing power in the hands of individual users. This posed problems for the corporate Information Technology hierarchy with duplication of files and complex systems for security and data integrity. "Thin clients" tried to redress some of these problems by distributing the processing power but centralising data storage. Until the advent of "cloud computing" the main focus has largely been around retaining processing power and storage on the desktop. There are many reasons why this has remained so both from an individual user perspective and a technology limitation perspective. Until cloud computing "became of age", the telecoms bandwidth, central processing power and data storage management were really inadequate to meet the demands of users for power and control.

Cloud computing is a real "game-changer" that brings a lot of centralised control and usage tracking back into society. This trend is likely to continue because we will value access to huge amounts of processing power and managed data storage that are charged on a transparent "per use" basis. Those who use these resources the most will pay the most but the power will essentially be available to all. In a sustainable society, this connection between usage and cost is essential and I believe it will come to be accepted even though users have been accustomed to services where the same monthly fee (or no fee at all) provides unlimited access to services such as the world wide web, electronic mail and internet telecommunications.

Cloud computing is also significant because it gives the hosts and providers of these services tremendous access to "the wisdom of the crowd" which in turn changes the relationship between citizens/users and technology. Our profiles, preferences and usage patterns enable these providers to deliver increasingly personalised services which grow our dependence on technology and change our relationship with technology from "mankind controlling machine" to "machine controlling mankind". It is my view that this trend could legitimately be described as mankind building a God in his own image and that all the attributes that we ascribe to a human-deity relationship will be mirrored in our attitudes towards and perception of this new era of cloud computing.

Imaging Technologies

Imaging technologies allow us to capture and record our physical world as still images and/or moving video. The technologies already available on digital cameras and even mobile phones provide very realistic and accurate representations of people, places and activities and the graphics display technologies provide us with a very acceptable quality of experience that comes close to "being there". There is no problem with "local" imaging and display technologies, even on mobile devices and the continuing development of broadband and wireless technologies will deliver the same or similar quality of experience streamed from "distant" locations.

Once the detectable quality of experience reaches its maximum, the focus will be on artificial intelligence that combines the expertise of the professional photographer/video producer with personalisation of imaging to the individual user's preferences. The creative tools use to manipulate and publish images and video will become ever more sophisticated, personalised and easy to use and cloud computing and social networking applications will empower us to share our creativity with ever wider audiences.

We are currently at the beginning of a 3D imaging revolution in which capturing and replaying images and video in three dimensions will begin to become mainstream. There are already consumer versions of 3D displays and cameras and these will develop in quality and ease of use. I envisage that the professional systems that are used by the entertainment industry will begin to enable users to create their own 3D perspective of movies.

Entertainment Technologies

Entertainment technologies have consistently migrated from public venues such as cinemas and clubs through the home to personal mobile devices with ever shortening timescales. The challenge for the owners and managers of public "out of home" entertainment centres is to provide sufficiently engaging experiences to attract customers away from what they have access to in the home or on the move. Cinemas are deploying very high resolution digital projectors, 3D imaging and high fidelity surround sound in order to deliver experiences that are not available to home or personal users.

Apart from home systems that can match the visual and audio quality of cinemas, the principal focus of entertainment technologies will be on mobile and portable devices such as phones and tablets. 3D capability and surround sound will be embedded in these devices but I expect to see advances in personalisation and user control/creativity on these platforms. The portable devices will "learn" the preferences and lifestyles of their owners/users and will suggest/prompt entertainment around these preferences. I anticipate also that collaboration / sharing technologies will become more sophisticated to allow consumers to share entertainment experiences with friends in a virtual way, initially in the home, but later on through portable devices.

Finally, the "prosumer" phenomenon will filter through into portable entertainment applications, allowing citizens/users to add their own creative and publish their own rich 3D immersive entertainment content.

Education Technologies

Education technologies are arguably the most significant development area and the one which will most impact the lives of all generations. The next few years will see revolutionary changes in practices at all levels of education. The so-called Generation Y of Digital Natives who were born after 1992 have never known a world without digital media communications and personal devices that provide communication, entertainment and education on demand. The richness of experience that these technologies provide creates a level of expectation for engaging activities that the traditional "Chalk and Talk" teaching methods just cannot match. One of the ironies of the demise of the knowledge professional as a hub for disseminating knowledge and skill is that the coach/facilitator/mentor/guide role replacing the traditional teacher role is probably a reflection of more ancient practices within tribes and older civilisations rather than a totally new concept. Educational technologies are taking us back to the future.

Good teachers have always been able understand and empathise with individual learners, tailoring their approach to deliver the most effective learning outcome for each student. Education technologies which once focused on improving the productivity of learning practices, are now more likely to focus on personalised learner-centric solutions that use artificial intelligence to adjust learning programmes to match the needs, expectations, preferences and capabilities of the individual. It is in education that personalisation will have the most significant impact on our development from cradle to grave.

The immersive technologies described in Chapter Nine use artificial intelligence to build profiles of technology users and this profiling helps devices and applications to better understand the interests, preferences, needs and capabilities of individuals in a much deeper and more intimate way. This growing body of knowledge about the individual user will be used not only to strengthen bonds between users and applications, but also to act as an e-portfolio for lifelong learning, recruitment, assessment and continuous development. Right from early years, learning in the home and nursery through primary, secondary and tertiary education converging and immersive technologies will embed formative assessment techniques not only to check learner progress but also to provide data for recruitment and career development. It is likely that learners, educators and human resource professionals will have increasingly sophisticated and personalised tools to support self-directed learning within communities of interest.

All this may mean that, almost from the time a person is born, they will have a personal profile of capabilities and preferences that could be used to create a personal learning and development path based on both their demonstrated skills and interests/passions. This is a highly contentious idea from a personal privacy perspective but there are already learner-centric programmes being developed in places like Korea that are based on similar principles and follow the "passions" of the learner. My own experience with my local milkman teaching himself web design skills in order to pursue an interest in his community is witness to the massively increased potential for learning where there is real motivation aligned to the interests of the learner.

Sensor and Location Technologies

Our advanced senses of sight, hearing, touch, taste, smell and "intuition" have been a significant differentiator between living creatures and machines. Now we are at the threshold of an age in which machines will be equipped with more powerful, sensitive and reliable sensors than human beings. We have already developed visual recognition technologies to read car number plates and verify passport photographs, sound recognition technologies that can convert speech to text and multi-sensor technologies for the "intelligent hospital bed" that can detect and diagnose serious illnesses and conditions. Our location technologies support our navigation along our roads and in our skies.

The implications of a technology infrastructure capable of tracking and monitoring activities and conditions and using the information to make judgements and take action are truly profound. Sensor and location technologies will go way beyond what George Orwell envisaged in his book "1984" about a "Big Brother" society because the tracking and monitoring aspects are likely to be less significant to individual rights and responsibilities than the artificial intelligence and "crowd-sourcing" techniques used to make judgements and initiate actions based on the sensor technology.

This is a massively important area for our sustainable and equitable future and one which will determine the balance between the rights and responsibilities of the citizen and the needs of society.

Serious Games and Immersive Technologies

Immersive technologies will continue to be at the vanguard of technological innovation as it is those applications which prove to be most successful at gaining and retaining our attention, time and money that will drive developments across all the sectors covered in this book. At the heart of all of these immersive applications will be the focus on building a highly personalised relationship with the user that enables each person to shape their own unique experience whilst belonging to a valued community, tribe or brand.

It is likely that sensor data will play an important part in this personalised relationship and the growth and strengthening of bonds that tie the end user to suppliers whom they trust to partner with in helping to shape their lives. One example of where this is likely to create a major but disruptive commercial opportunity is in health care. Most people aspire to being fitter and healthier but are insufficiently motivated to accept any pain or sacrifice in their daily lives to adopt the kind of sustainable lifestyle that could make any significant difference. This is compounded by the lack of real time monitoring and feedback on the health parameters that reflect general health.

The growing availability in retail stores of blood pressure and body mass index monitors is an indication that if people can be motivated to make daily use of such health data and at the same time have easy access to products and services designed to improve health, this could make a real difference to the general health of the population. I see this being developed through immersive technologies that take health monitoring data that is captured in an ambient way through sensors embedded in clothing, watches, jewellery etc., and use this data to provide personal coaching not only through on-line mentoring within social networks but also through artificial intelligence linked to that sensor data in exercise machines or so-called exergames. The personal data could be stored on "the cloud" to make it accessible to the user anytime and anywhere and also to provide a personal health database that could be "entrusted" to a chosen healthcare/insurance provider.

The personal health sector is one of the most obvious examples where sensor data and artificial intelligence could be combined to provide new portfolios of personalised services that would dramatically change the landscape of existing commercial service provision. Other sectors where the opportunity to develop and use immersive technologies in this way could be energy usage (through smart meters), education, learning and development (through smart games-based learning) and recruitment. Development of market sectors in this way could help to bring some stability and sustainability to society because it would help to redress some of the disruptive disconnections in all kinds of relationships that have emerged over recent years.

Past the Tipping Point

Once the internet was "invented" in the 1990s by Tim Berners-Lee, the world changed irreversibly and the widespread accessibility of knowledge and publishing tools made the acceleration of change and innovation inevitable. The implications of this tipping point are very profound. We have reached the end of a beginning which started with the transfer of our intelligence and uniquely human capabilities into machines. We have already reached the point where the intelligence that can be contained in and/or accessed by a portable device arguably exceeds that of a normal human being. We may be on the final leg of a journey where the destination is unknown and completely unpredictable.

The Beginning of the End?

I regard myself and all my fellow baby boomers to be the most fortunate of any generation ever to have existed. Born into a world rising like a Phoenix from the ashes of the Second World War, I have never been called to fight in foreign lands like my father and grandfather and never lived in fear of my best years being snatched away in the service of my country. My grandfather was captured as a seventeen year old messenger boy in the First World War's killing fields of France and forced to work as a prisoner of war in the coal mines of the Ruhr but he was one of the lucky ones who came home to their families in 1918. My father was with the RAF in India in the Second World War and reading his letters home from India and his diary notes written on board the troop ship which carried him back to England is a very moving experience that makes me reflect that whilst his teenage years witnessed hardship and death, mine were spent in the pursuit of hedonistic pleasures and the exploration of the beautiful world we live in. As if a lifetime free from the hardships and conflicts of war were not enough, I have also witnessed the emergence of technologies that are smashing the dimensional constraints of space, time and state to open up a world of opportunities undreamt of in previous generations. These have been the best of times that could yet still be the worst of times if we prove to be incapable of properly and sustainably harnessing all the power and intelligence of the technologies we have developed.

I always believe that it is better to travel hopefully than to arrive at a final destination and I certainly feel that my life's journey through these amazing technological developments is far from over. Every human being with access to technology is choosing this journey's path into the future through the time, attention and money they spend on specific applications. Today we choose to invest most of our discretionary time, attention and money on activities which engage, motivate, entertain and educate us and create opportunities for us to shape our lives and the world we live in. These so-called immersive applications of video games, virtual worlds and social networks are our personal gateways to the future that drive the direction of research and development into new devices and innovative applications. The question is whether the choices we make today about our relationship with technology will leave us no choices in the future. In our efforts to master technology will we only succeed in becoming its slave?

The next steps in mankind's journey into the future will be critical and like any journey worth planning, we need to reflect on where we have come from, the places we have visited on our journey, what lessons we might draw from those excursions and what we hope to gain along our future path.

Where to From Here?

This chapter is intended to be a controversial wake up call. I want the previous chapters to be like a comfy familiar armchair that you drift into a contented nostalgic sleep in before being awoken by an alarm to discover that the world around you has changed beyond recognition and you wonder how you can survive in this new and alien landscape. The previous chapters have sought to show how, in the span of one man's lifetime, the human race has begun an irreversible process of transferring all its intelligence and knowledge of good and evil into the machines that we humans previously needed all our skills and intelligence to get the best from. Today it is these machines and the technologies we are developing that are using their senses and intelligence to get the best out of us human beings. I have witnessed first-hand during my business life this transition not only through the emergence of "Prosumer" tools eroding the need for human experience and skills but especially also during my time under the control of sophisticated logistics technologies in the frozen food warehouse. The question is "What kind of world are we creating and will this path lead to the liberation of the human race or its enslavement or, worse still, its destruction.

Here are a few radical thoughts on what the future holds:

1. God is being created in the likeness of man
2. The Old Testament is not about the past but about our future
3. Have we already invented time travel?
4. Treating modern day society like a nuclear reactor
5. The most connected society in mankind's history is also the least connected
6. The way ahead has to be the assignment of individual IP addresses to every human being on the planet and a new concept for human rights and individual freedom.
7. The essential nature of the paradox in our existence – Yin and Yang
8. Immortality within our grasp

1. God created in the likeness of man

My faith in an ultimate deity has been a rock all through my life and will continue to be so. There is no tangible evidence that I can point to that proves beyond doubt that my God exists but I know in my heart that he does live all around us. As I look at the characteristics of the converging technologies described in this book, I see a changing relationship between humans and technology that is beginning to mirror our relationship with God. When I used to say the Lord's Prayer before I went to sleep I shared my innermost feelings and fears with God. In church I would be told from the Bible that we should give 10% of our earnings to God.

Today in cloud computing and the various devices we all use, we are effectively increasingly transferring our faith and our innermost thoughts to a new Deity whose memory, intelligence and senses are spread around the globe and to whom we look for our daily bread. This new God has been created in our own likeness and we willingly give a good deal more than 10% of our most precious resource, time, as well as large amounts of our discretionary income. Where we once relied on the power of prayer and human friendship to find our life's partner, increasingly we are now finding our wives, girl-friends, husbands and boy-friends through the matching skills of internet dating sites.

I believe that our current use of immersive technologies is driven by a fundamental human need for personal relationships which value us as individuals, comfort us in times of need, motivate us to higher aspirations and provide a bedrock on which to base our lives. I can see evidence all around that what was once provided by our faith in a Higher Being is now being entrusted to our personal gateway to immersive technology applications.

2. The Old Testament is not about the past but about our future

I have always assumed that the Bible's Old Testament was about our historical past, not meant to be taken literally, but loosely based on mankind's past mistakes and slow path to the salvation outlined in the New Testament. What if this is the wrong way round and that the Old Testament is about the future ? Was I born into the metaphorical Garden of Eden where we could commune freely with God and are we about to be cast out because we have tasted the apple from the tree of life which brought knowledge of all good and evil. Is the internet the apple that the serpent offered to us all. It is perhaps somewhat ironic that I wrote the best part of this book in my own Garden of Eden paradise, the Domaine de Belezy naturist camp, where I and my fellow holiday makers are all able to walk about naked and unashamed.

In the Old Testament the world made its way to salvation with the ethos that the path to heaven was through leading a good life and the message of "an eye for an eye and a tooth for a tooth" is perhaps a profound statement about sustainability being based on actions and reactions and rights and responsibilities instead of our world today where technology makes it possible for someone on one side of the planet to profoundly affect the life of another human being without any consequence to themselves.

In my own personal interpretation of the Old and New Testaments, I see a world in the Old Testament in which there was a straightforward relationship between doing good deeds and going to heaven and doing wrong and being punished. Simply put, there was a philosophy of "an eye for an eye and a tooth for a tooth" and a direct relationship between cause and effect, virtue and rewards and evil and justice. This mirrors our experience of the physical world but makes no allowance for the creative and unpredictable human spirit. In contrast, the New Testament suggests that a genuine belief in Jesus Christ is sufficient to secure a passage into heaven and the essential connection between cause and effect, virtue and reward, evil and punishment is severed.

The machines of my childhood were definitely of the physical world, predictable in nature with a strong link between cause and effect. Today's machines with cloud computing, ubiquitous sensors, crowd-sourcing and artificial intelligence are increasingly much more ethereal and unpredictable

3. Have we already invented time travel?

I am not so arrogant as to suggest that the scientists who can prove it is impossible to travel faster than light and therefore also impossible to travel through time are wrong. My premise has nothing to do with the commonly held perception of time travel as being like Doctor Who or H G Wells. It is more a philosophical argument that says that we can look at the concept of time travel differently.

There was a philosophical question once posed along the lines "If a tree falls in a forest where there are no human beings to observe, did the tree really fall ?". This is an interesting argument that assumes that change is only apparent if it is observed by a human being but the argument is fallacious because, whether observed or not, the fallen tree does impact its environment and over time, rotting in the ground, may provide the fuel that changes someone's life in the future.

We all make the observation about time flying or dragging as it seems to our senses to be capable of going slowly or rushing by. I do not believe that we will ever be able to physically go back in history and take actions that alter our future but I do believe that our digital technologies are now making it increasingly possible for our past history and our future projections to alter the present in ways that have never been possible before except on a very limited scale.

In some of my presentations on immersive technologies I often show an ancient cave painting dated at around 10,000 B.C. and showing the hunting of the creatures of that time. Those paintings undoubtedly influenced the cave dwellers of the time but for many thousands of years they lay unseen and undiscovered, contributing nothing to the current shape and understanding of the time. Today they have been photographed, digitised and posted in books and on the internet for many thousands of people to see and learn about and although the influence of seeing an ancient painting may be minimal for any individual human being, collectively they have begun once again to alter human perceptions and actions in the present and future. Now our digital archiving and our historical research capabilities are resurrecting and preserving the past in newer, richer and immersive ways that can be witnessed by the whole world with unpredictable consequences. Similarly, our forecasting techniques can bring the future into the present and affect the outcomes. If this exponential trend continues to its singularity, all of our past and all of our future will converge at a single point in time where time reaches its maximum perceived speed and then stops.

4. Treating modern day society like a nuclear reactor

The internet and atomic energy have certain things in common. They are man-made discoveries which can be used for good or evil and now that they have been invented there is no going back – we cannot un-invent either of them. The difference between nuclear energy and the internet is that our global society prevents proliferation of nuclear energy to stop it falling into the wrong hands whereas we have made the power of the internet accessible to virtually everyone on the planet regardless of their potential usage for good or evil.

I have only a layman's understanding of atomic energy but I do know that this fantastic power has been harnessed to generate light and heat peacefully and that mankind has developed ways of controlling the energy through cooling and managing the forces within the reactor. Nuclear fission happens when there are no such controls and an uncontrollable chain reaction is triggered with devastating results.

It is somewhat timely and ironic that I read an extract from a book called "the Fear Index" written by Robert Harris, whose background is in nuclear physics. In this extract published in the UK's "Daily Mail" newspaper he suggests that a number of nuclear physicists, who were made redundant when a massive USA project called the "Desertron" (the equivalent of CERN's Hadron Collider) ran out of funding, found employment on the Wall Street Stock Exchange where their skills were redeployed to develop sophisticated automated investment programmes which use artificial intelligence to buy and sell stock, making vast fortunes for a small group of very rich and elite investors, but threatening the stability of the global financial markets. He ends this extract of his book with the words:

> *"It is that the financial system itself has somehow slipped all human control – that it has become the preserve of the profoundly anti-democratic, super-rich elite, and that it girdles the planet like some alien entity from an H.G. Wells novel. The digitised financial machine does not work for us; we work for the machine and I do not believe that our political leaders have the faintest idea how to bring it under control."*

My question is whether the internet has triggered the equivalent of a societal chain reaction in which the massive volume of interactions causes exponentially increasingly unstable consequences. If this analogy is valid, how can we put into place the equivalent cooling mechanisms to harness this power for good?

5. The most connected society in mankind's history is also the least connected

I am an engineer by profession and am familiar with processes that obey the laws of the physical universe. I believe that Isaac Newton once observed that he could predict the courses of the stars and planets but not the actions of a mad man. In my final year at Birmingham University I built an analogue computer with a closed loop feedback system which connected the output of the computer into the processing system and effectively controlled the computer to respond to the results of its actions. Our brain is programmed in a similar way and it is how we learn to make sense of the world we live in and manage our actions based on our experience of the consequences of what we do or think. A child who puts his fingers too near the fire will hopefully learn that such an action is not worthwhile. There is a clear connection between action and consequence. It was largely so throughout history and human beings in the role of mediators, facilitators and middle-men fulfilled a role that brought sustainability and substance to all kinds of relationships in all sectors of life. For most of history, every citizen, except a privileged few, has had very limited ability to connect to or influence other people.

Today, the internet and mobile telecommunications has put the power of global communications into the hands of everyone without the responsibility for the consequences of their actions. Power without responsibility is a recipe for total disaster whether that power is in the hands of a dictator or a beggar. I have lost count of the number of people that I have never met yet have subsequently learnt that they have been influenced by my words or actions without me ever knowing. Sometimes these connections bring about unexpected positive results but very often, as today's tabloid media will witness, they can have negative unforeseen consequences. This kind of situation is totally unsustainable and needs to be acted on quickly before it is too late.

6. The way ahead has to be the assignment of individual IP addresses to every human being on the planet and a new concept for human rights and individual freedom

There will always need to be a balance between human rights, civil liberties and the needs of society to protect and preserve itself. The situation today, where on the one hand we demand freedom from being observed and monitored by the increasing number of cameras and sensor devices all around us, and on the other hand the basic societal need to protect itself from actions by people who accept no responsibility for those actions is potentially untenable.

Almost all computers have a transactional logging capability built-in to be able to "roll-back" or at the very least track activities. An increasing number of devices are now having embedded biometrics and/or face recognition built into the security and logging-on routines and in my view these developments do need to be pursued vigorously to redress some of the consequences of the "disconnected" connectedness of the internet.

Our freedom of choice and the preservation of society could both be protected by drawing up a new charter of civil rights and assigning every human being with a unique IP address that can irrefutably be linked to their true identity. In this charter, every human who "walks in the light" and is totally transparent in their thoughts and deeds would be rewarded by unrestrained access to all mankind's communications services. Those individuals who wish to keep all or part of their lives secret would be able to do this provided they lost the unrestrained access to communications. This need not be a permanent arrangement for either choice and citizens could choose their moments of privacy and openness as the need arises but society would be protected this way.

7. The Essential Nature of the Paradox in our existence – Yin and Yang

This notion is that all living things exist because of the impossibility of their existence. In the physical world, we are used to the concept of a state or condition and the notion that any entity cannot simultaneously have opposing conditions e.g. an object cannot be hot and cold at the same time. Matter and anti-matter cannot co-exist. We cannot simultaneously live in the best of times and the worst of times.

My thought is that the paradox is an essential part of our existence and that opposites have to co-exist like Yin and Yang in Asian Philosophy. I would go further to suggest that when we reach the extremes of a condition, we actually achieve exactly the opposite of that condition, e.g. Ultimate freedom is ultimate slavery (Man is free but everywhere in chains).

In this view of the world, the path to leadership is through following, and the path to ultimate success is through total failure.

I believe the paradox is what differentiates human beings from machines. Modern technologies are already processing, accessing and storing data more effectively than the human brain. Machines are now capable of "learning" and "remembering" in more powerful ways. Sensor technologies have begun to overtake the sensitivities of human sight, hearing, touch, taste and smell. Artificial intelligence and crowd sourcing are endowing computers with "human-like" judgement and perception but human beings have a life force, "soul" and "raison d'etre" that machines will never acquire.

8. Immortality Within our Grasp

Immortality has been a Holy Grail for mankind across the thousands of years of our existence and now mankind has the potential to become immortal in more than one sense. Many of the people that I have met and whose works are described in my book are already immortal in a very real sense. Just as Christianity's Jesus Christ and the great Prophets of other religions still influence the attitudes and behaviours of countless millions of people, the works and thoughts of prominent men and women have been archived in digital format to be discovered and accessed for future generations and their contribution to mankind will continue as long as humans and computers exist side by side on our planet.

When I had the task of clearing out my mother's home before she moved into care, I discovered many long-forgotten and, in some cases, unknown memorabilia in the form of photographs, letters and diaries that belonged to members of my family that are no longer with us. When I look at the faded smiles on these sepia coloured prints and cuttings I am transported back in time to their feelings and their emotions and memories and their presence is with me once more – as long as such items exists, as increasingly they will, these people will remain immortal and ready to make their mark on all who discover their legacy.

In the physical sense also, developments in genome technology and our growing ability to implant technology inside human beings could see the span of active and functional life extended substantially and potentially forever. Finally, through the creation of our alter ego avatars we have made representations of ourselves that could continue to live on and function through artificial intelligence indefinitely. Our avatars do not yet age or decline and, on the contrary, actually have the potential to continue to learn and develop long after their human creator has left this world.

This book raises many more questions than answers but I firmly believe that we need to find answers to those questions very soon or the better world we all wish to see future generations inherit will remain an illusion.

ABOUT THE AUTHOR

David Wortley at work on this book in Domaine de Belezy, Provence, France

David Wortley is a freelance consultant on the strategic use of immersive and emerging technologies such as serious games, virtual worlds and social networks. He is a Fellow of the Royal Society of Arts (FRSA) with a career that embraced the converging and emerging technologies of telecommunications (Post Office Telecommunications), computing (IBM), digital media and community informatics (Mass Mitec, a rural SME) and the creative industries (De Montfort University Leicester, UK). He is a serial entrepreneur and innovator with a passion for applying technology to social and economic development

From 2006 to 2011, David was Founding Director of the Serious Games Institute (SGI) www.seriousgamesinstitute.co.uk at Coventry University and was responsible for the development of the Institute as a global thought leader on the application of immersive technologies (which include video games; virtual worlds and social networking) to serious social and economic issues such as education; simulation; health; commerce and climate change. Working with academics; regional development agencies and leading computer games companies, David made the SGI a focal point for games based learning, simulation and immersive 3D virtual environments and an engine for innovation and social and economic regeneration.

David is a respected (see http://davidwortley.com/testimonials.html) and sought-after international conference speaker and writer for global publications on Learning Technologies, Defence and Health applications. He has written numerous papers on technology and society (see *http://www.davidwortley.com/articles.html*) and is a regular conference presenter (see *http://www.davidwortley.com/conferences.html*

INDEX

'Ere be Dragons, 325
2001 Space Odyssey, 168
35mm, 34, 75, 77, 80, 143, 160-162, 168, 169, 171, 173, 175, 177-183, 186, 190, 243
35mm Express, 75, 162, 163, 168, 169, 243
3-D, 114, 173, 229, 233, 260, 288, 298, 301, 303, 321, 323, 339, 341, 342, 358
5120 - IBM 5120 Computer, 127
84 Charing Cross Road (Book, Film and Play), 21

Aardman Animations, 261
Abanazar, 222
Acme Clothing Company, 132
Action Maze (Serious Game), 67, 242
actions and reactions, 350
Active Ingredient, 324
Addlesee, 23
Adobe, 111, 114, 260
Advantage West Midlands, 327, 328
AEI, 28, 210
Aeroflot, 254
African Queen, 53
Ageing Society, 283
Agilent Technologies, 90, 92, 109, 250, 251
Aherne, Mike, 230
Aidensfield, 106
Airfix, 6, 123
Airplane, 182
Aken, Dick van, 89, 91, 189

Akulova, Alice, 256
Aladdin, 21, 222
Alesund, Norway, 36
Alfie (Lunchtime business meeting), 52
Alfreton, Derbyshire, 182
Alien, 230
Allan, Alex, 87, 88, 264
Alright Now (Song by Free), 204
Alsop, Cliff, 121, 126
AMARC, 101
Amazon, 87, 264, 275
Ambient Performance, 329
America's Army, 331, 332
Ames, John, 34, 61
Amidon, Debra, 99, 112, 275
Amiga, 183
Amsterdam, The Netherlands, 249
Amstrad, 73, 144, 145, 216, 224
  Lotus 1-2-3, 141
Andrews, Dana, 203
Animals, The, 199
Annecy, France, 252
Annoni, Alessandro, 298
APL (Advanced Programminmg Language), 127
Apollo, 154, 289
Apple, 74, 141, 167, 175, 217, 243, 284
Appleby Magna, Leicestershire, 23
Applied Research, 329, 330
Archer, Jeffrey, 318
Archway, London, 67
Arnold, P.P., 206
Artificial Intelligence, 53, 265, 283,

338, 341, 343, 344, 345, 350, 352, 356
Artois, Rene, 222
Ashbourne, Derbyshire, 29
Askew, Robin, 61
Assembler, 119
Astle, Mr, 62
Astronomy Domine (by Pink Floyd), 28, 201
asynchronous, 57, 73, 245
Atari, 320
Atkins, Madeleine, 302, 330
ATM, 273, 300
Atom Heart Mother (by Pink Floyd), 28
Atomic Energy Authority, 181
Atomic Rooster, 199
Auckland, New Zealand, 48, 114, 287, 326
augmented reality, 339
Austin, Carolyn, 162
Australia, 113, 299
Autocad, 159
Autonomy, 246
avatars, 328, 356
Ayutthaya, Thailand, 50, 51
Azimuth Co-ordinator, 201

baby boomer, ix, 1, 54, 265, 346, 309
Back to the Future, 229
Backhouse, John, 30, 31
Bad Girls (TV Series), 232
Bahadur, Lachse, 98
Bakelite, 59, 115, 191
Banbury, 254
Bandrova, Temenoujka, 277, 298
Bangor Maine, USA, 47
Barcelona, Spain, 102, 133
Barco, 169, 171

Barcovision, 169
Bardney, Lincolnshire, 16
Barker, Ritchie, 210, 319
Barker, Ronnie, 121, 149
Barnsley, South Yorkshire, 38, 210, 230
barometer, 268
Barraclough, Dave, 316
Barratt, Sid, 201, 206, 208
Barton, Dave, 185
Basingstoke, Hampshire, 137
Bath, Somerset, 26, 28
BBC, 60, 86, 193, 196, 197, 232, 242, 243, 266
Beano (Children's Comic), 193
Beasley, Nadine, 224
Beast from the Black Lagoon, 203
Beast Mart, 239
Beatles, The, 26, 29, 206
Beauty and the Beast (Pantomime), 224
Beckham, David, 307, 309
Bed Full of Foreigners (Comedy Play), 212
Bedford, Craig (Uncle), 28, 56, 57
Bee Gees, 117
Beeching Axe, 16
Beijing, 294, 297, 299
Belfast, Northern Ireland, 180
Belize, Naturist Resort, France, 100, 357
Benjamin Franklin School of Global Education, 93
Bentine, Michael, 196, 197
Berchtesgaden, Germany, 249, 250
Bergen, Norway, 36
Berkeley University, California USA, 245, 287, 288, 291
Berlin, Germany, 32, 103, 273

INDEX

Bernard Shaw, George, 1, 221, 224
Berners-Lee, Tim, 346
Berry, Sue, 143, 162, 165
Bert Weedon's Play in a Day, 204
Bertie the cat, 165, 172
BETT, 258, 259
Beverly, Humberside, 44, 46
Bexhill, East Sussex, 67, 119, 241, 322
Big Blue, 128
Big Brother, 344
Big Wheel, 320
Bigfern, 84, 261
Biggles, 238, 266
Billy Cotton Band Show, 193
Bingley, Frank, 85, 261
biometrics, 339, 354
Birmingham, West Midslands, 17, 20, 34, 36, 37, 61, 62, 110, 118, 119, 143, 158, 160, 177, 206-209, 232, 236, 241, 259, 315, 316, 353
Birstall, Leicester, 19
Biryani, 120
Black Five, 18
Black, Cilla, 203
Blackberry, 281, 339
Blackboard, 261
Blackwall tunnel, London, 67
Blake, Amie (Grand-daughter), 177
Blake, Carol (Second Wife), 46, 47, 75, 76, 95, 174, 175, 223
Blake, Christopher (Stepson), 46, 50, 175, 177, 178
Blake, Stephen (Stepson), 46, 109, 177, 178
Bletchley, Bedfordshire, 31, 117
Blisworth, Northamptonshire, 51, 52
Blithe Spirit, 224
Blitz Games, 113, 164, 260, 328

Blue Remembered Hills (Play by Dennis Potter), 231
Bluebird (unexpected sale), 122
Blues Brothers, 66
Bluetooth, 53, 54, 110, 337
Bluto, 194
Bob a job, 311
Bodicoat, Mark, 218
Bogart, Humphrey, 53
Bolton Wanderers, 307
Bomb, The, 4, 5, 21, 22, 269
Bonanza, 195
Bonzo Dog Doodah Band, 192
Boothbys, The, 8
Bootle, Lancashire, 29, 60
Boots, 72, 158, 200
Borrill, Cyril, 308
Boston, Lincolnshire, 2, 3, 9, 12, 15, 16, 18, 21-34, 40-42, 49, 50, 57-61, 79, 99, 112, 157, 158, 173-175, 191-205, 209, 211, 237, 238, 240, 266, 270, 306-314, 318, 320, 335
Boston United, 100, 195, 306, 308, 313, 314
Bowie, David, 261
Bowley, Denise, 23
Bowser's playing field, 194
Box Brownie, 157
BRADS, 127, 133
Braid, 70, 71, 142
Brammers, The, 8
Branson, Richard, 209
Branston, Julian, 215
Breeze, Alan, 193
Brick Bradford, 194, 202, 320
Bridges, Bob, 85, 86, 107, 225
Bridges, Lloyd, 182
Briggs, James, 109

Brighton, Sussex, 195, 255
Bristol, 126, 128, 261
Britannia Class, 2, 3, 10
British Council, 102, 109
British Telecom, 65, 244, 247
British Waterways, 95
Brixworth, Northamptonshire, 34
Brno, Czech Republic, 286-289
broadband,ix, 90, 92, 247, 321, 331, 337, 341
Brooke Bond, 238
Brooks, Keith, 130
Brooks, Kevin, 166
Brown, Gordon, 101
Browns, The, 8
Bruce, Jack, 204
Brussels, Belgium, 92
BT, 28, 64, 65, 72, 92, 144, 334
Buckby, Daniel, 223
Buenos Aires, Argentina, 92
Bumblies, 197
bum-flap, 132
Burdasses, The, 8
Burden, David, 329
Burgess, Sid, 270
Burroughs (Computer Supplier), 68, 135, 136
Burton, Richard, 231
Bushy Rovers, 308
Buzby, 65

Cake Walk, 320
Cakir, Ahmet, 249
Calcomp, 171-175
Calders and Grandidge, 7
California, USA, 167, 218, 285, 287, 288
Cambridge, Cambridgeshire, 6, 8, 240, 246

Campbell, Robert, 87, 88
Canada, 41, 46, 137, 248, 260, 294, 322
Canals, 12, 50, 51, 53, 95, 221
Candida (Play by George Bernard Shaw), 224
Canon, 176, 177, 188, 208
Capewell, Gary, 80, 163, 167
Capewell, Ray, 164
Careful with the Axe Eugene (by Pink Floyd), 207
Carnaby Street, London, 26
Casper The Ghost (Musical), 223
Caspian Learning, 332
Caterpillar, 320
Caucus, 84, 245, 246
Cawston, Jacqueline, 257, 330
CD-ROM, 144, 177, 244, 254
Ceefax, 73, 244
cerebral palsy, 101, 114
CERN, 352
Cessna, 40, 44
chalk and talk, 240, 266, 343
Chandler, Peter, 263, 323
Channel 2020, 325
Channel Islands, 49
Chapel en le Frith, Derbyshire, 18, 311
Chapel St Leonards, Lincolnshire, 23, 24, 50
Cherry Corner, 306
Chicken Korma, 120
Chinks again (Marbles Game), 310
Chinley, Derbyshire, 18
Chips Away, 45
Chopra, Deepak, 112, 275
Chris Sharpe Motorcycles, 144
Cisco, 115, 247, 334
civilisation,ix, 3, 12, 105, 343

INDEX

Clapton, Eric, 207
CLC 700, Canon Colour Copier, 188
Cleese, John, 121, 128, 134
Click and Create, 187, 322
Clickteam, 323
Cliff Richard, 219
Clifton, Brian, 312
cloud computing, 156, 297, 340, 341, 349, 350
Clough, Brian, 120
Clun Castle, 18, 19
Cockayne, Steve, 85-88
Cockerill, Richard, 231
Codeglia, Ken, 90
Codemasters, 113, 260, 323, 328
Colin, The Geordie, 33
collaboration technologies, 82, 84, 89, 93, 247, 248, 263
Comknet (Community Commerce and Knowledge Network), 84-89, 92, 94, 100, 103, 108, 111, 246, 249, 251, 286
Commodore Pet, 243
communications technologies, 2, 75, 85, 89, 92, 173, 243
Community Informatics, 92, 105, 107, 108, 113, 189, 286
Compaq, 144, 170, 171
Compuserve, 82
Computing Technology, 117-156
Concorde, 41, 42
Coningsby, Lincolnshire, 33, 41
Connery, Sean, 216
Consett, County Durham, 102
Cooden Beach, East Sussex, 67, 119, 139, 241, 242
Cook, Thomas, 13-16, 42, 54, 86, 88, 93, 108, 223, 291
Cook, John Mason, 15

Cooney, Ray, 225
Copenhagen, Denmark, 33, 49, 273
Corbett, Harry, 60, 197
Corbett, Ronnie, 121
Corby, Northamptonshire, 84
Corby Glen, Lincolnshire, 5
Corel, 167, 169, 174, 184, 187, 255, 322
Corona, 21
Coronation Scot, 195
Coronation Street, 304
Corti, Kevin, 263
Courtaulds, 71, 72, 139, 143
Coventry, Warwickshire, 101, 111, 114, 138, 288, 289, 302, 326-330, 357
CP Programming Services, 133
CPSR (Computer Professionals for Social Responsibility), 277
Crazy World of Arthur Brown, 199
Cream (Rock Group), 199, 204
Creative Industries Knowledge Network, 110, 259, 260, 261, 323
Cromer, Norfolk, 311
Cromwell Road, Great Glen, Leicester, 38, 211
Cruickshank, Iona, 111, 114, 259-261
Crystal Palace Bowl, London, 201, 220
Crystals, The, 199
Curse of the Werewolf, 214
Curve Theatre, Leicester, 162, 263
Cushing, Peter, 202
CVS (Council for Voluntary Services), 94

Daft, Joan, 228

Dainty, Gordon, 211
Dalby, Tyrone, 204, 205
Dale, Jim, 167
Damai Sciences, 277, 283
Dan Air, 38
Dandy, 193
Dansette, 198, 201
Dark Side of the Moon (by Pink Floyd), 215
Darling Buds of May (Book, Play and TV Series), 8, 221, 222
Darlington, County Durham, 80, 178, 179
Dassault Systemes, 334
Datsun Sunny, 54
Daventry, Northamptonshire, 109, 147, 153-155
Dawkins, Ian, 135
Day, Graham, 218
de Freitas, Sara, 329, 330
De Montfort University, Leicester, 48, 65, 110, 111, 113, 154, 187, 258-260, 283, 287, 301, 322-326, 357
Death of a Salesman (Play by Arthur Miller), 221
DEC (Digital Equipment Corporation), 135, 252
Defence Global, 263
Del Boy, 194
Dell, 154-156
Delray Beach, Florida, USA, 46, 47
Delta, 142
Depot, The, 323, 324
Derby, Derbyshire, 16, 28, 29, 176, 183, 195, 259, 307, 314, 325, 326
Derbyshire, UK, 13, 18, 53, 139, 238, 266

Desertron, 352
Dhaka Deli, 120
Dhupa, Venu, 109
Dickens, Charles, 238, 336
diesel, 16, 17, 51, 53
Diggins, Ron, 198
Diggola, 198
Digital Earth, 48, 277, 282, 285-288, 291-298, 326
Digital Immigrants, ix
Digital Natives, ix,2, 235, 265, 342
Dingley, Northamptonshire, 37, 210
Dire Straits, 217
Dirty Dozen, 63, 150
Disco Dave, 259
Diskfax, 77, 78, 80, 178- 180
Disney, 46, 53, 224
Disneyland, Florida, USA, 168, 229
Displaywriter, 121
disruptive technologies, 335
DLP – Digital Light Processing, 185
DMS - Diploma in Management Studies, 65, 213
DMU - De Montfort University, 113-115, 258-262, 323-326
Doctor Terrors House of Horrors, 203
Doctor Zhivago, 151
Dollis Hill, London, 30, 31, 60, 159
Doncaster, South Yorkshire, 12
Doonreay, Scotland, 181
Dore, 18
Dorothy, Aunt, 56
DOS - Disk Operating System, 75, 141, 154, 169, 175
Dougiamas, Martin, 254
Downing Street, London, 88
DPD – Data Processing Division (IBM), 121, 122, 137

INDEX

Dracula, 203, 211
Dream Team (TV Series), 232, 233
Dubbin, 308
Dublin, Eire, 180
Dulcie Triumph Herald DUL 814C (my first car), 36-38

East Sussex, UK, 67, 68
Eastenders, 86, 304
Eastleigh, Hampshire, 146
Eccles (from The Goons), 197
Edinburgh, Scotland, 5, 12, 17
Edmondson, Bud, 216
Educational Technology, 235-266
Ehlers, Manfred, 277, 293
EI7, 63, 209
Ektachrome, 161
elearning, 255, 258
Eliot-Higgit, Mike, 328
Emerson, Keith, 200
encryption, 79, 117, 281, 339
End of chute, 148
End of the Beginning or Beginning of the End, 335-356
Enderby, Leicester, 206
engagement, 290, 293, 304, 305, 311, 321, 322, 330
Engelbart, Douglas (Inventor of the Computer Mouse), 244, 284, 289, 289
Enigma Code, 31
Entertainment Technology, 191-234
Entovation, 99, 112, 275
Entwhistle, John, 204
Epcot, Florida, USA, 47, 229
Equals, The, 199
ERDE – European Rural Development by means of Education, 102, 107, 251

ERNIE, 272
eRooms, 251
Ever Ready, 193
eVideo, 88-90, 248
Executive Engineer, 62, 63, 66, 209
Execuvision, 143, 160-162, 243, 322
exergames, 345
Extras, 232

Facebook, 56, 160
Fairford, 42
Fairy Dell, 23
Family Rock Tree, 199
Famous Five, 238
Fawlty Towers, 133, 317
Fawlty, Basil, 128, 317
Feather, Pauline, 35
Fenside Road, Boston, 193, 237
Ferguson, Alex, 128
Ferranti Mark I, 117, 118
Ferrari, 175
Fiat, 37, 53, 54, 300, 338
film recorder, 143, 163, 169, 173, 180, 188
Finch, Ian, 35, 317
Finney, Tom, 307
Firsby, Lincolnshire, 16
First Cut is the Deepest (by PP Arnold), 206
Fisher, Don, 34, 61
Fishtoft Creek, Boston, Lincolnshire, 99
Fisons Pharmaceuticals, 72, 170
Flash Gordon, 194
Fleetwood Mac, 33
Flight of the Bumble Bee, 192
Floreat Bostona, 100, 239
Florida, USA, 46, 47, 229
Flower power, 26

Flying Scotsman, 5, 17, 20
Follow the Man from Cooks, 13, 223
football, 2, 15, 26, 35, 37, 62, 64, 100, 193, 194, 206, 210, 305-308, 311, 314-319, 321
Foresman, Tim, 277, 286-289
Fork Lift, 148
Fortran, 118, 119, 284
Forward Trust Group, 177
Foundation for Water Research, 109, 110, 183-187, 323
Foxton, Leicestershire, 52, 221
Frampton Marsh, Boston, Lincolnshire, 23, 49
France, 50, 139, 230, 252, 346, 357
Francis, John, 290, 291
Fraser, Andy, 204
Free (Rock Band), 204
Freelance, Lotus, 75, 169, 174, 183, 243, 322
Freiburg, Germany, 35
Fujix, 188
Fuller, Buckminster, 285, 286, 291
funfair, 195, 233
FWR - See Foundation for Water Research

Gadget Show Live, 197, 242
Galley, Jim, 126
games based learning, 114, 263, 357
gamification, 331
Gang Show, 191, 196, 211, 270
Gardner, Ava, 196
Garfitts Lane, Boston, Lincolnshire, 308
Garforth, Darren, 231
Gargery, Joe (Great Expectations), 234
Gates, Bill, 141, 154

Gatwick Airport, London, 47
Geesin, Ron, 208
Gell, Alan, 185
Genesys, 98
geospatial, 264
Germany, 35, 97, 100, 102, 249, 278, 283, 293, 295, 317
Gerry and the Pacemakers, 203
Gervais, Ricky, 232
Ghengis Khan, 117, 150
Ghost train, 320
Gilberts, The, 7
Gilmour, David, 206, 207
Gimson Road, Leicester, 37, 210
Gipsy Bridge, Boston, Lincolnshire, 29, 62, 314, 315
Glasgow, Scotland, 181
Glastonbury, Wiltshire, 26
GLD - Global Learn Day, 93
Gledhill, Joe, 241
Gliderdrome, 26, 27, 158, 199, 202
Global Marshall Plan, 279
Global Society Dialogue, 97, 100, 103, 277-279, 282, 286
Global Villages, 103
Global Watches, 130
Glover, Brian, 230
Goathland, North Yorkshire, 34, 106, 229
God, ix, 9, 156, 270, 341, 348, 349
Godber, John, 230
Golden Wonder, 78, 188
Golders Green, 270
Goldfinger, 216
golf ball typewriters, 121
Goodmans, 206
Goods In, 148, 152
Google, 248, 261, 264, 291, 292, 301, 327

Goon Show, 196
Gopel, Maja, 281
Gore, Al,x, 279, 285, 298
Gotham City, 205, 211
Gott, Clive, 49
GPO, 28, 59, 62-65
GPS – Global Positioning System, 51, 281, 301, 302, 325, 339
Grammar School, 23, 32, 37, 56, 99, 196, 198, 205, 210, 238-240, 308
gramophone, 4, 5, 191, 198
Grantham, Lincolnshire, 11, 12, 16, 18, 314
Granville Street, Boston, Lincolnshire, 237
Great Central Railway, 18, 19
Great Expectations, 238
Great Glen, Leicestershire, 38, 40, 44, 66, 120, 131, 210, 216
Grey, Tom, 211, 212
Grimsby, Lincolnshire, 2, 5, 6, 16, 34
Grundtvig, 102
Guitar Hero, 304
Gulliford, Mick, 29
GWR – Great Western Region, 18

Hadley, John, 131, 132
Hadron Collider, 352
Hall, Doug, 121, 133
Hall, Richard, 261
Hamilton, Jacque, 227, 228
Hammer Horror, 202, 215
Hancocks, Johnny, 307
Hanspal, Sonia, 166
Harborough and Bowden Charity, 95
Harborough Community Learning Network (HCLN), 93-98, 103, 105, 108, 109, 111, 138, 146, 251, 275, 286
Harborough FM, 98, 99
Harborough Theatre, Market Harborough, 53, 167, 217, 220, 223, 224
Harrier Jump Jet, 41
Harris, Bill, 89-91
Harris, Robert, 352
Harry Potter, 86, 107, 229
Harvard Graphics, 75, 78, 146, 169, 174, 179- 183, 229, 243, 322
Hatton, Phil, 133
Hawkwind, 209
Hayes Modems, 73, 74, 81
Haymarket Theatre, Leicester, 13, 217, 223, 231, 263
HCLN – See Harborough Community Learning Network
Head, Anthony, 219
Health Global, 263
Heartbeat, 34, 106
Heathrow Airport, London, 42, 46, 51, 131, 138, 145, 153, 169, 254, 258
Heaven is a place on Earth, 183
Hedblom, Milda, 98, 279
Hedgehoppers Anonymous, 217
Hednesford, West Midlands, 319
Heidelberg, Germany, 35, 317
Heighway, Steve, 316
Helander, Martin (Professor), 249, 250, 277, 283
Helga (Film), 203
Helsinki, Finland, 33, 49, 273
Helter Skelter, 320
Henderson, Tom, 218
Hendrix, Jimi, 28, 34, 209
Hepburn, Kathryn, 53
Hercules Graphics Card, 142

Hernando's Hideaway (from Pyjama
    Game), 214
Hewlett Packard, 90, 135
Hey there, you with the stars in your
    eyes (from Pyjama Game), 214
Hibbs, John, 89, 93
High Bay, 148-153
High Hall,University of
    Birmingham, 35, 205-208
Hill, Ken, 214
Hilversum, 4
Hinckley, Leicestershire, 65, 177,
    210, 319
Hipkins, The, 7
His Master's Voice, 192
history of mankind, 236
HMV – His Masters Voice, 192
Hobbit, Village in Poland, 104
Hodby, Barry, 211
Hodges, Brett, 178
Hogg Robinson Group, 138
Hoggard, Hobson, 113, 328
Hogwarts, 107, 229
Holloway Road, London, 67
Homepride, 189
Hopkins, Anthony, 21
Horncastle, Lincolnshire, 16, 315
Horwood House, Little Horwood,
    Buckinghamshire, 31, 60
Hotspur (Children's Comic), 193,
    307
Howe, Steve, 52
Howshams, The, 76, 77
HP 3000, 137, 172
HP Media Solutions, 90-92
HPVC – HP Virtual Classroom, 90,
    92
HTML – Hypertext Markup
    Language, 85

Huadong, Guo, 293, 296
Hubert (Ski Instructor), 39
Hughes, Marian, 66
Hukin, Arthur, 307
human beings,ix, 1, 3, 9, 11, 109,
    116, 117, 119, 135, 149, 153,
    155, 184, 222, 235, 236, 246,
    264, 267, 269, 270-274, 276,
    282, 284-286, 290, 292, 321,
    340, 344, 348, 350, 353, 355
Human Computer Interface, 282, 284
Humber Snipe, 5, 22, 53, 194
Humberside, North Yorkshire, 45
Humerdinck, Englebert, 38
Hunchback of Notre Dame, 214
Hungary, 102
Hunstanton, Norfolk, 5, 22, 23, 196,
    272, 320
Huntingdon, Cambridgeshire, 6
Hurricane, Hawker, fighter plane, 41
Huthwaite Research Group, 125
hypertext, 244, 290

I can't get no satisfaction (by Rolling
    Stones), 208
IBM, 46, 68-71, 77, 91, 95, 120-137,
    139-144, 154-156, 159, 160,
    164, 168, 169, 173, 179, 183,
    222, 225, 243, 252, 273, 284,
    322, 357
ICANN, 97, 278
ICL (International Computers Ltd),
    135-137
ICT – Information Communications
    Technology, 99, 100, 105, 107,
    277
identity, 104, 148, 273, 274, 300,
    339, 354
Idziak, Waclav, 103

Imaging Technology, 75, 78, 80, 114, 143, 157-190, 293, 301, 339, 341, 342
immersion, 233, 317, 321, 330
immersive activities, 306, 309, 317, 321, 335
Immersive Technologies,ix, 48, 114, 223, 235, 254, 263, 292,303-334, 338, 343, 345, 349, 351, 357
in-flight information system, 48
Incheon, South Korea, 107
India, 6, 56, 98, 107, 114, 205, 263, 268, 346
InFocus DLP Data projector, 185
Infoterra, 301
Innit, 87, 93
Instamatic, 160
Intel, 82, 247, 333, 334
intelligence,ix, 4, 11, 37, 53, 54, 55, 57, 75, 109, 117, 119, 135, 146, 147, 149, 154, 156, 184, 189, 190, 237, 264, 266, 268, 272-276, 290, 304, 346-349, 355
internet,ix, 12, 52, 56, 73, 81-90, 94, 97, 106, 108, 110, 115, 159, 245, 246, 254, 261- 264, 273, 274, 281, 282, 322, 330, 335, 337, 340, 346, 349-354
Ioannidis, Spiros, 187
Ion, Canon still video camera, 176, 177
IP address, 354
ISDE – International Symposium for Digital Earth, 47, 277, 286, 287, 292, 293, 297, 299
ISDN – Integrated Services Digital Network, 81, 82, 247
Isle of Wight, UK, 26

Istance, Howell, 323
It's good news week (by Hedgehoppers Anonymous), 217
ITEC, 165
Ivanhoe (HMS), 49, 318

James Went building, Leicester Polytechnic, 65
Japan, 90, 108, 287, 294, 320
Jecks, Clive, 46, 137
Jeffrey, Dick, 129, 130, 131
Jerwood, Henry, 328
Jesus Christ, 219, 350, 355
Jevremovic, Alex, 329
Johannesberg, South Africa, 52, 89
John Bull Printing Set, 308
Johnny Remember me (by John Leyton), 198
Jones, Arthur, 217
Jones, Jenny, 42
Jones's Laughing Record, 192
Joule, Ian, 220
Jules Verne Astroglide, 127, 230

Kaftan, 26
Kanyeihamba, Ruth, 178
Keegan, Kevin, 100
Keet, Sarah, 189
Kennedy Memorial Test, 317, 319
Kermode, John, 138
Kes, 230
Kettering, Northamptonshire, 52, 163
Khaikina, Alla, 253
Khalid, Halimahtun (Professor), 249, 277, 283
Kibworth, Leicestershire, 14, 68, 139, 165, 166, 211-218
Killer Ants from Space, 203

Kinect, Microsoft, 303, 324
Kings Cross, London, 2, 6, 12, 15, 16, 17, 60
Kitwood Boys School, Boston, Lincolnshire, 23, 238
Knowle House, Market Harborough, 94
Knowledge Ecology Network (KEN), 245, 251
Kodak, 157, 158, 161, 176
Kompressomatic 3000, 255
Konecny, Milan, 97, 277, 279, 286, 298
Korea, South, 107, 344
Kremlin, The, Moscow 257
Krypton Computers, 135
Kuala Lumpur, Malaysia, 250, 277
Kyoto, Japan, 90, 248
Kyunju, South Korea, 107

Labelsco, 135
Lambretta, 25, 26, 33, 50, 150
LAN - Local Area Network, 81, 82
Lancaster, Avro, 4, 6, 41, 268
Langar Airfield, Nottinghamshire, 42- 44
Langford, Bill (Uncle), 4, 8, 22, 23, 49, 194, 195, 306, 318
Langford (Grandma), 2, 5, 173, 192-194, 198
Langford, Iris (Aunt), 4, 23
Langford, Len (Uncle), 18, 19, 21, 41, 81, 146, 306, 309
Langford, Mick & Steve (Cousins), 194, 195
Langford, Pop (Grandad), 2, 8, 17, 192-193, 222, 223, 306, 309, 346
Langham, Roy, 76

Large, Paul, 232
Larkin, Pop & Ma, 221
Laserjet, HP, 177
Last of the Red Hot Lovers (Play by Neil Simon), 222
Last Tango in Whitby, 228
Lawrence,D.H., 224
LDS – Leicester Drama Society – See Little Theatre
Leamington Spa, Warwickshire, 113
Leapfrogging, 117, 150
Learning Management Systems, 254
learning technologies, 103, 236, 237, 242, 252
Lee, Christopher, 202
Leicester, 13-15, 19, 20, 28, 36, 37, 40, 42, 62-71, 82, 84, 94, 110, 111, 114, 119, 130, 137-139, 147, 158, 162, 165-167, 177, 182, 206, 209, 210, 213, 214, 217, 221, 223, 225, 227, 231, 258, 259, 262, 263, 301, 319, 323-325, 357
Leicester Tigers, 231
Leicestershire, 5, 41, 52, 87, 96, 97, 98, 120, 139
Leningrad, USSR, 32, 33, 49, 273
Lennons, The, 8
Lester, Diann, 66, 319
Lester, Malcolm, 66, 214
Leyton, John, 198
Li150 (Lambretta), 25-29, 150
Licari, Dan, 328
Lichfield, Staffordshire, 44
Liggins, Stephanie, 231
Lillehammer, Norway, 36
Lilleshall, Shropshire, 24
Lincoln, Lincolnshire, UK, 16, 28, 121

INDEX

Lincolnshire, 2, 12, 18, 19, 21, 24, 26, 99, 121, 127, 155, 194, 204, 252, 314, 317, 318, 335
Linden Labs, 113
Lisbon, Portugal, 170
Lithuania, 102-104
Little Horwood, 31
Little Theatre, Leicester, 139, 191, 212, 221, 222, 225, 226, 227
Liverpool, Lancashire, 29, 193, 305, 316
LNER – London North Eastern Railway, 4, 11, 17, 19
Loading, 148, 151, 152
location, 51, 52, 68, 77, 78, 83, 96, 104, 105, 107, 110, 132, 139, 149, 168, 176, 189, 205, 211, 232, 271, 272, 275, 277, 283, 288, 292, 299-302, 310, 324, 344
Loman, Willie, 221, 222
London, 2, 4, 7, 8, 14-21, 28, 30, 32, 33, 42, 47, 48, 51, 59, 60, 61, 67, 69, 71, 72, 74, 86, 87, 92, 101, 102, 109, 114, 126, 129, 139, 144, 159, 163, 166, 168, 170, 174, 177, 186, 195, 219, 221, 223, 225, 226, 232, 233, 246, 249, 258, 263, 270, 305, 311, 329
Lone Ranger, 195
Loot (Play by Joe Orton), 226, 227
Lord of the Rings, 87, 104
Loughborough, Leicestershire, 14, 19, 20, 37, 84, 123, 170, 210, 316, 319
Loughborough Colleges, 316
Louth, Lincolnshire, 252, 317
Love City, 325
Low Brow Trash, 324

Lubenham, Leicestershire, 75, 76, 89, 111, 147, 165, 178, 185, 188, 220, 260
Luft, Tim, 330
Lumpkin (from She Stoops to Conquer), 218
Luton, Bedfordshire, 38
Lyon, France, 35

Mackay, Celia, 225
Mackay, Richard, 138, 139, 225
Maelstrom, 324
Maid Marian Way, Nottingham, 121
Main, Douglas, 227
mainframe, 121, 135, 137, 159, 340
Malaysia, 249, 283
Malechova, Poland, 103
Malkinson, 27
Mallard, 5, 11
Mallyan Spout Hotel, 106
Malmesbury, Wiltshire, 110
Manchester, 117, 128, 181, 305, 307
Manchester United, 128, 307
mankind,ix, 1, 3, 10, 11, 50, 116, 155-157, 184, 236, 245, 264-267, 271, 272, 277, 289, 291, 294, 303, 310, 330, 336, 341, 347-349, 352-355
Mannesmann Tally, 70
Manuel (from the TV Comedy Fawlty Towers), 133
marbles, 310
Marconi, 28
Market Deeping, Lincolnshire, 18
Market Harborough, Leicestershire, 13, 14, 37, 52, 56, 75, 78, 84-89, 95-97, 126, 138, 163, 167, 181, 188, 210, 216-220, 223-225, 249, 261

Marlow, Buckinghamshire, 110, 183, 184
Marvin, Lee, 63
Mass Mitec, 69-75, 78, 81-84, 86, 92-98, 115, 139, 140, 143, 144, 162-166, 169, 173-181, 184, 186-190, 220, 243, 260, 321, 322, 357
Massachusetts, USA, 99, 174
Matthews, Mary, 328
Mayflower, The, 3, 99
McGowan, Ian, 120
McGuigan, Barry, 92
Meadow View, Leicester, 72, 73, 163, 165, 172
Meccano, 6
Melbourne, Derbyshire, 13
Meredith Vula, Lala, 262
Mersey Sound, 206
MESI – Moscow State University for Economics, Statistics and Informatics, 252-253
Messerschmidt, 22, 126
Messiter, Malcolm, 74
Mezdunarodnaya (Moscow Hotel), 257
MHDS - Market Harborough Drama Society, 167, 218-224
Miami, Florioda, USA, 46-50
Mickey Mouse Club, 194, 320
Microlight, 44
Microsoft, 74, 90, 92, 111, 141, 146, 154, 169, 174, 175, 183, 187, 238, 243, 248, 303, 322
Middlesbrough, Teesside, 61, 102, 105, 106
Midnight Hour (by Otis Redding), 204
milkman, 21, 85, 146, 318, 344

Miller, Arthur, 221, 222
Milligan, Spike, 196
Milton Keynes, Buckinghamshire, 31, 60, 185
Milton, Ken, 225, 227
MIRACLE, 283
Mitchell, Edgar Apollo 14 Astronaut), 289
Mitsurina, Italy, 38
MM5C Ltd, 96
MMDP –(Multimedia Demonstrator Program), 83
mobile devices, 339-342
modem, 69, 73, 74, 77, 81, 90, 110, 116
Mods, 25
Moira, Leicestershire, 23
Moldova, 98, 279
Monkhouse, Bob, 179
Montage film recorder, 173, 174
Montreal, Canada, 46, 137
Moodle, 254
Moore, Jeanne, 221- 224
Moore, John, 225, 226
Moore, Rebecca, 292
Moore, Sally Ann,x, 252
Moore, Siobhan, 223
Moore's Law, 337
moped, 24
Moran, Gerard, 111, 259
Moreton, Thame, Oxfordshire, 110
Mornington Crescent (from Radio Quiz Show), 74
Morris, Mike, 257
Moscow, Russia, 32, 33, 49, 153, 233, 251-256, 258, 273
Mosquito, De Havilland, 42
Mothers Club, Erdington, Birmingham, 207

motivation, 78, 108, 178, 246, 264, 304, 311, 315, 317, 321, 330, 344
Motorola, 70
Move, The, 199
MP3, 54, 217
Multimate, 70, 71, 141
Munro, Andy, 211
Munro, Janet, 211
Muscat, Oman, 112, 113, 275
Music Hall, 213, 220, 227
My Generation (by The Who), 204
My Guy (by Mary Wells), 198

Nahrada, Franz, 102
Narborough Road, Leicester, 120
NASA, 127
National Girobank, 144
National Presentation Network, 78, 80, 179, 183
NCR – National Cash Registers Computer Supplier, 68, 120, 135, 136
Neath, Debbie, 221
Neill, Tim, 255, 256
Neilson, Leslie, 182
Nepal, 98, 279
Nessebar, Bulgaria, 298
NESTA, National Endowment for Science, Technology and the Arts, 109
NetMeeting, Microsoft, 248
New Testament, 349, 350
New Theatre Cinema, Boston, 192
New York, USA, 91, 167
New Zealand, 47, 114, 287, 295, 326
Newbery, John, 123, 127
Newbold Revell, Leicestershire, 66, 68

Newmarket (card game), 309
Newton, Isaac, 353
Nextspace, 288
Nice (Rock Group), 29, 139, 200, 206
Nighy, Bill, 198
NME (New Musical Express), 209
North Hykeham, Lincolnshire, 121
North Kilworth, Leicestershire, 139
Northampton, 131, 132, 186
Northamptonshire, 34, 51, 98, 163, 167
Norway, 36
Norwich, Norfolk, 240, 314
Not Now Darling, 213
Nottingham, 15, 16, 19, 29, 33, 42, 109, 120, 121, 124, 130, 133, 135, 137, 139, 165, 169, 246, 324
Nottinghamshire, 43

O'Grady, Dari, 24
O'Malley, John, 225
Oadby, Leicestershire, 42, 69, 70-72, 75, 139, 161, 163, 165, 171
Oakham, Leicestershire, 28
Oboe Fantasia, 74
Odeon, The Cinema, 26, 192, 194, 320
Oh Calcutta, 219
Old School House, Market Harborough, 95
Old Testament, ix, 348, 349, 350
Olive Oyle, 194
Oliver Cromwell, 2, 10, 20
Olivetti, 128, 129
Olympia, London, 166, 258
Oman, 112, 275, 325
Only Fools and Horses (TV Series), 194

OPD – Office Products Division (IBM), 121, 128
Optical Laser Disk, 241
Opus (Multimedia Software), 254
Orlando, Florida, USA, 46, 47, 229
Orton, Joe, 226, 227
Orwell, George, 344
Ower, Jude, 328
Oxford,UK, 71, 177, 240
Oxford, Ken, 314
Oxfordshire, 52, 110, 181

P&P Micros, 144
Paddington, London, 17
Page, Jenny, 229
Paintshop, 255
Palette, 143, 160-163, 169, 173, 190
Palin, Michael, 108, 113
Pannell Kerr Forster, 165
Panther Kalista, 175
parachute, 42, 44
paradox, 348, 354, 355
Parikh, Jyoti, 98, 279
Paris, France, 36
Parkes, Alison, 224
Parlophone, 192
Parry, Ian, 218
Parry, Martine, 114, 263
Pathe, 194
PC Anywhere, 74
PC, IBM, 140
PC World, 146, 189
PCM - Pulse Code Modulation, 64
PDA (Personal Digital Assistant), 325
Peak Practice, 232
Pebble Mill, Birmingham, 232
Peel, John, 34, 208
Pegasus, 47

Pelling, Dennis, 44, 45, 231
Pender, David, 112, 325
Penelope (Canal Boat), 52
Penny Lane, Liverpool, 29
Perrin, 241
Perry, Ann, 212
personalisation, 49, 53, 87, 265, 337, 338, 341-343
Perth, Australia, 299
Peter and Gordon (Pop Group), 28
Peterborough, Cambridgeshire, 2, 5, 6, 11, 16, 61, 195
Phantom of the Opera, 223
Phillips, 144, 244
Phoenix Theatre Leicester, 214, 262, 346
Phones 4U, 256
Photo CD, 176
Pickering, North Yorkshire, 46
Pickett, Horace and Nellie, 21, 63
Pictography, Fujix, 188
Pilgrim Fathers, 3, 99
Pink Floyd, The, 26-30, 34, 117, 200, 201, 204, 206, 207, 209, 215, 220
Piper at the Gates of Dawn (by Pink Floyd), 200, 201
Pitney, Gene, 203
PixeLearning, 263, 328
Placeware, 92
Plague of the Zombies, 203
Planetwalker, 290
Playgen, 114, 263
Plymouth, Devon, 99
Points of View, 60
Poland, 32, 102-104
Polaroid, 143, 160-163, 169, 173, 175, 190
Ponderosa, 195

INDEX

Pong, 320
Popeye, 4, 194
Por, George, 84, 245
Porridge, 149
Porter, John, 69, 139
Portland, USA, 89, 260
Portugal, 170
Post Office, 28, 29, 56, 59, 60- 68, 106, 110, 119, 120, 153, 159, 205, 241, 242, 247, 319, 322, 357
Post Office Telecommunications, 28, 65, 66, 110, 119, 159, 205, 241, 357
POTS, 93
Potter, Rob, 262, 325
Powell, Michael, 113, 260, 261, 323
Powerpoint, Microsoft, 75, 78, 82, 88, 99, 101, 111, 146, 169, 174, 179, 180, 183, 243, 248, 322
Prague, Czech Republic, 186, 187, 322
predictability, 269, 272
Presley, Elvis, 191
Prestel, 73, 244
Price, Vincent, 203
Private Walker, 194
Proby,P.J., 199
projector,data, 34, 106, 146, 169-171, 185, 186, 244
Prontaprint, 78-82, 142, 145, 178-183, 189
Proshare, Intel, 82, 248
Prosumer, 336, 342, 348
Provence, France, 252, 357
punched cards, 118, 284
Pyjama Game (Musical Play), 214

Quantel Paintbox, 143, 160
Quicksilver Girl (by Steve Miller Band), 270
Quintar 1080, 169, 173

R G Mitchell, 126
Radermacher, Franz Josef (Professor), 97, 277, 278
radio, 4, 12, 34, 63, 64, 74, 87, 91, 94, 98, 100, 101, 153, 172, 193, 196, 198, 205, 208, 216, 233, 249, 251, 264, 269, 275, 300, 307, 320
Radio Caroline, 198
Radio Interference, 63
Radio Luxemburg, 198
Radio with Pictures Show, 97, 98, 101, 109, 112
RAF – Royal Air Force, 6, 33, 40, 42, 56, 178, 205, 217, 268, 346
Ragg, John, 226, 227
railways, 2, 3, 7, 11- 13, 16, 18, 21, 108, 195, 237, 306
Raleigh, Bicycles, 23, 24, 33
Rbase, 71, 142, 145, 246
Rcader, Ralph, 270
Real Towell, 315
Red Arrows, 41
Red Square, Moscow, 32, 257
Redding, Otis, 199, 204
Reeves, Christopher, 217, 225
Regal cinema, Boston, Lincolnshire, 192, 202, 203
Reisensburg, Germany, 97, 278
relationship,ix, 3, 4, 9, 14, 22, 24, 34-37, 56, 95, 112, 113, 133, 139, 171, 183, 186, 242, 251, 268, 274, 336, 340, 345-350
Rentacrowd, 39, 42, 66, 69, 139, 162, 215, 320

responsibility, 35, 63, 65, 129, 159, 214, 280, 353, 354
Retford, Nottinghamshire, 12
Rheingold, Howard, 107
Richardson, Martin, 114, 260
Richmond, North Yorkshire, 106
Richmond Events, 49
Rickard, Di, 222
Riddett, Alec, 261
Riddett, Dave Alex, 261
Rigby, Norman, 307
Righteous Brothers, 199
rights and responsibilities, 281, 338, 344, 350
Riley, Al, 316
River Deep Mountain High, 158
River Haven, Boston Lincolnshire, 23
River Welland, 23
roadshows, 181, 182, 203
Robert Smyth Grammar School, Market Harborough, 37, 38, 84, 210
Robin Hood, 2, 6
Robinson, Bob, 135
Robinson, Robert, 60
Rockers, 25
Rockingham Road, Market Harborough, 96
Rolling Stones, The, 26, 208
Rolls Royce, 28, 176, 316
Rolodex, 141
Roma Nova Virtual World, 332
Roman, 12
Ronettes, The, 199
Ross, Paul, 92
Roundtree, Graham, 231
Rowbotham, Lynette, 212
RPG (Report Programme Generator), 71, 133
Run for your Wife (Play by Ray Cooney), 225, 226, 227, 231
Russell Square, London, 60
Ruthin Castle, North Wales, 174
Rutland Street, Leicester, 323

S/1 - IBM Series 1, 135
S/32 - IBM System 32, 122, 123, 124, 125, 134, 135, 155, 156
S/34 - IBM System 34, 122, 123, 133, 134, 135, 136, 137, 138, 139, 140, 142, 143
S/36 - IBM System 36, 132, 133, 134, 136
S/38 - IBM System 38, 136, 137, 138, 142
Sachs, Andrew, 133
Saemoul, 108
sage on the stage, 266
Salmon, Dave, 63
Salzburg, Austria, 249, 250
Sam the Sham and the Pharoes, 217
San Diego, California, USA, 89, 104
San Francisco, California, USA, 168, 245, 289, 290
Sanderson, Marion, 211
Sands, Gilbert, 314
Sarawack, Malaysia, 249
Saucerful of Secrets (by Pink Floyd), 207
Scaffold, The, 28
Scarborough, North Yorkshire, 229
Schauer, Thomas, 98, 100, 277, 279
Schuler, Doug, 278
scouts, 7, 18, 21, 196, 311, 312, 313
Scrupps, The, 8
Seagoon, Neddy (from The Goons Comedy Radio Show) 197

INDEX

377

Searle's (camp site), 22
Season's Greetings (Play by Alan Ayckbourn), 228
Seattle, USA, 245, 278
Secombe, Harry, 196
Second Life, 113, 260, 302, 330
Sekonda, 130
Sellars, Peter, 196
Sensor and Interface technologies, 267-303
sensors, 339, 344, 345, 350
Seoul, South Korea, 107, 108
Serbia, 262
Sergeant Pepper, 206
Sergeant Porterhouse, 225
Serious Games, 48, 67, 113-115, 154, 155, 187, 252, 254, 257, 258, 260, 263, 288, 289, 292, 302, 305, 309, 310, 315, 322, 326-332, 334, 344, 357
Serious Games Institute, 48, 113, 114, 115, 154, 155, 257, 260, 263, 288, 289, 302, 326, 327, 328, 329, 357
Set the Controls for the Heart of the Sun (by Pink Floyd), 28, 207
SFX, 107, 168, 225
SGI (see Serious Games Institute)
Shankly, Bill, 305
Shaw-Parker, David, 223
She Stoops to Conquer (Play by Oliver Goldsmith), 217
Sheffield, South Yorkshire, 15, 16, 19, 38, 182, 207, 210, 319
Shodfriars Lane (Boston United Football Ground), 306
Sibsey, Lincolnshire, 33, 61
Siggraph, 166, 167, 218
Silver Machine (by Hawkwind), 209

Simon, Neil, 222
Simpson, Richard, 277, 288, 298
Sinclair, 164, 322
Singapore, Asia, 48, 249
Sixsmith, Ian, 40
Skegness, Lincolnshire, 5, 6, 8, 16, 18, 23, 25, 61, 126, 127, 133, 196, 230, 314, 317, 318, 320
skiing, 38, 273
Skype, 248
Sleaford, Lincolnshire, 16, 18, 121
slot machines, 127, 192, 196
Slough, Buckinghamshire, 88, 89
Small, Keith, 96
Smart, Linda, 232
SME – Small to Medium Enterprises, 324, 357
Smellaround, 214, 215
Smith, John, 63
Smith, Rachel, 46, 80, 163, 168
Smith, Roy, 225, 227
SMS – Short Message Service, 57, 325
Snade, Geoff, 307, 312
soccer, 24, 35, 65, 100, 148, 193, 210, 239, 305, 308, 315, 319
social networking, ix, 247, 248, 264, 281, 305, 308, 310, 315, 328, 331, 320, 334, 335, 339, 341, 347, 357
Softeach, 145, 168, 169, 175
Softsel, 144, 145, 168, 174
Solden, Austria, 39
Solitaire film recorder, 180, 188
Sooty, 60, 197
Sorokin, Sid, 214
South Kilburn, London, 87, 93
South Korea, 107, 108
South Leicestershire College, 96
Southampton, Hampshire, 49, 81

space and time, 11, 270-272, 292, 306
Space Invaders, 187, 217, 320
Space Mountain, 229
Spaceship Earth, 285, 287, 291
Spalding, Lincolnshire, 18, 21, 222, 314
SPC - Software Publishing, 168, 169, 174, 175
Spector, Phil, 158, 198, 199
Spike Jones and the City Slickers, 5, 192
SPIN – Situation, Problem, Implication, Need, 125
Spinks, David John, x
Spinks, Jenny, 25, 42, 118, 199, 202
Spitfire, Supermarine fighter plane, 4, 41
Sporting Club Bicker, Lincolnshire, 315
Squash Leicester, 66, 217, 320
St Anton, Austria, 39
St James, Melanie, 289, 293
St Petersberg, Russia, 32
St Trinians, 231
St Tropez, France, 50
Stafford, Ray, 174
Staniland Primary School, 202, 237, 238, 266, 308
Stansted Airport, London, 47, 249
Star Inn, Sibsey Lincolnshire, 33, 61
Star, Kam, 114, 263
Starlight Express, 223
Stay with me Baby (by Sharon Tandy), 207
steam, 3-6, 11- 20, 28, 32, 46, 53, 167
Stephen, Caryll, 109, 110, 183-185
Steve Miller Band, 270

Stevens, Simon, 101, 114
Stixwold, Lincolnshire, 16
Stockley, Bob, 317
Stokes, Alf, 60
Stone, Staffordshire, 205, 319
Stoneygate, Leicester, 38, 210
Stoneygate Road, Leicester, 38
Stoughton, Leicester, 40, 41, 44
Strawbs, The, 200
Strowger, 59, 61, 63
Sudbury Towers, London, 126
Summerfields, The, 7
Sunderland, County Durham, 33
Superman, 1, 168, 217, 225
Sutherland, Ian, 68
Swan National, 137
Swansea University, South Wales, 316
Swinging Sixties, 25, 26
Sykes, Eric, 225
Symington, William, 14
synchronous, 57, 245
System X, 28, 59

T Rex, 199
Tabirtsa, Stella, 98, 279
Taito, 320
Tandy, Sharon, 207
Targa, 176
Tarrant, Chris, 60
Taylor, Rick, 66
Taylor, Simon, 320
TDM - Time Division Multiplexing, 64
Teamstation, 82, 247
Teesside, 105-107
Telecommunications, 28, 56-116, 119, 143, 178, 263, 329, 337, 340, 353, 357

INDEX

telephony, 12, 54, 339
Telepresence, 115, 247
television, 2, 4, 5, 12, 32, 64, 77, 91, 92, 94, 114, 121, 131, 149, 164, 172, 177, 192, 194, 195, 199, 202, 203, 205, 210, 211, 217, 226, 228, 229, 232, 240, 244, 247, 251, 260, 320
telex, 64, 70, 71, 115, 142, 143, 171, 273
Tesco, 109, 147-150, 258
Thailand,Asia, 50
Thaisim, 50
The Bill (Television Series), 232
The day I saved the barge (Music Hall Monologue), 221
The Madcap Laughs(by Sid Barratt), 208
The Price (Play by Arthur Miller), 222
The Progressive Underground and Rock Society, 208
Thematic Villages, 103
Thinking Worlds, 332
Thompson, Jeremy, 218
Thompson, Steve, 101, 102, 105
Thomson data projectors, 171, 244
Thorntons, 182
Three Mills Studios, London, 233
Three Swans, Hotel and Restaurant, Market Harborough, 225
Tiger the cat, 172, 173
Tikhomorov, Vladimir, 153, 252, 253
time, space and state, 267, 272, 275, 277, 285
Timeline of Media Technology Adoption, 58
Timeline IBM General Systems Computers, 122

Timson, Dave, 39, 139, 214, 215
TNA (Tim Neill Associates), 255
Toft, Elaine, 228
Tonto, 195
Top Town Quiz (Anglia Television), 240
Toronto, Canada, 46, 137
Toshiba, 110
Tottenham Court Road, London, 30
Tower of Terror, 229
Townsend, Pete, 28, 204
Training Zone, 263
Transend, 74
transparencies, 75, 80, 171, 172, 181, 241
Transport, ix, 10-55, 83, 163, 178, 272, 291, 330, 337, 338
Travelsphere, 95, 138
Triumph Herald, 36, 54
Tromans, 32
Troubador, 218
Truscott, Jim, 227
Turley, Ian, 66, 320
Turner, Paul, 329
Turner, Tina, 158, 199
Twitter, 56
Two Way Family Favourites (BBC Radio Programme), 193
typewriter, 57, 64, 79
Tyseley, Birmingham, 17, 20
Tytton Lane, Wyberton, Boston, Lincolnshire, 3, 21, 270

Ulm, Germany, 97, 98, 278, 279
Ummagumma (by Pink Floyd), 207
Unichem, 69
Universal Studios, 168
Unix, 77
unpredictability, 4, 269, 272

Up n Under (Play by John Godber), 230
USA, 48, 50, 53, 88-92, 98, 99, 107, 112, 160, 166, 174, 260, 273, 294, 301, 352

Van Morrison, 203
Vancouver, Canada, 46, 137, 138
Ventura Publisher, 172
Venture Business Forms, 69, 71, 139, 142, 143, 163, 164
Vickerage, Ian, 109
video, ix, 48, 68, 81, 82, 86, 88- 90, 94, 107, 111, 113, 115, 164, 176, 177, 183, 185, 186, 217, 241, 243, 245, 247, 260-262, 273, 290, 303, 320, 322, 328, 331-333, 337, 339, 341, 347, 357
video games, ix, 48, 113, 303, 320, 322, 328, 331, 333, 337, 339, 347, 357
Vienna, Austria, 102, 103, 280
Vilnius, Lithuania, 104
Virgin, 46, 47, 209
virtual classroom, 90-92, 98-101, 109, 112, 248-251, 260, 263, 334
Virtual Pub Quiz, 106
virtual worlds, ix, 173, 293, 302, 305, 310, 315, 328, 329, 331, 333, 334, 339, 347, 357
Virtuality Ltd, 324
Vulcan, Avro, 41

Waddle, Chris, 319
Wakeman, Rick, 200
Wakey Wakey, 193
Wang, 135
Wanlip, Leicester, 65, 66, 319
Warley, West Midlands, 316

Warwick University, Coventry, Warwickshire, 316
Washington, USA, 48, 60, 199, 328
Watch with Mother (BBC Children's TV Programme), 197, 242
Waters, Roger, 204
Watkiss, Adam, 226, 228
Watson, Lynette, 225, 231
wax thermal, 172
Wayne Fontana and the Mindbenders, 203
Web 2.0, 255, 321
Webber, Barry, 189
webcam, 82, 90, 94
webcast, 88-90, 93, 98
webinar, 89, 111, 115
Wehsener, Solvig, 280
Wellington, Vickers, 6, 123
Wells, Angie, 165, 178
Wells, H G, 350
Wells, Mary, 198
West Runton, Norfolk, UK, 311
WH Smith, 200
Wharfedale, 205
What every woman knows (Play by George Bernard Shaw), 221
Whist, 196, 309
Whitby, North Yorkshire, 228, 229
Whitcombe, Lisa, 222, 224
White House, 75, 76, 78, 89, 94, 95, 98, 115, 147, 173-175, 185, 188, 220
White, Nancy, 245
White, Phil, 128, 129
Who, Doctor (TV Series), 86, 196, 350
Who, The, 25, 28, 117, 204
Wicks, Clive, 238
Widowing of Mrs Holroyd (Play by

INDEX

DE H Lawrence), 224
Wigfull, Tom, 206
Wiggington, Sue, 75
Wigston, Leicester, 42, 96, 97, 165, 166, 172, 226
Wildman, Des, 207
Williams, Claire, 51-52
Williams, Steve, 52
Williams, Kenneth, 227
Wilson, Heather, 211, 213
Wilson, Richard, 66, 320
Winchester, Hampshire, 81
Winchester Disk Drive, 124
Windows, Microsoft, 74, 85, 141, 169, 175, 183
wisdom of the crowd, 340
Witham, River, Boston, Lincolnshire, 16
Wizard of Oz (by Frank L Baum), 85
Woburn Abbey (Pop Festival), Woburn, Bedfordshire, 27
Wold Farm Foods, 123, 130
Wolverhampton Wanderers, 307
Woodgate, Peter, 298
Woodhall Spa, Lincolnshire, 16, 21, 311
Woodstock (Pop Festival), USA, 26
Woolworths, 29, 198, 200
Wooly Bully (by Sam the Sham and the Pharoes), 217
Workers Playtime, 193
Worraker, Willie, 23, 24
Wortley, Angela, 37, 38, 66, 111, 131, 158, 210, 211
Wortley, Betty Mabel (Mum), 2, 7, 23, 24, 38, 40, 41, 42, 43, 58, 62, 79, 81, 157, 182, 193, 194, 201, 237, 238, 268
Wortley, Grandma and Grandad, 237

Wortley, John Leslie (Dad), x, 2-7, 23-26, 35, 40, 41, 42, 57, 58, 79, 157, 193, 194, 201, 205, 208, 209, 237, 268, 307, 309, 311, 346
Wrangle, Lincolnshire, 314
Wright, Rick, 201, 208
WWDU – Ergonomics Association, 249, 250, 283
Wyberton, near Boston Lincolnshire, 2, 3, 9, 19, 58, 193, 198, 204, 237, 238, 268, 308, 310, 313-318

Xerox, 135

Yarborough Road,Wyberton, Boston, Lincolnshire, 4, 7, 8, 194, 311
Yardbirds, The, 199, 207
Yin and Yang, 348, 354
YMCA – Young Men's Christian Association, 30
YOIS – Youth for Inter-Generational Justice and Sustainability, 281
YOP - Youth Opportunities Scheme, 165
York, UK, 12, 91, 96, 166, 167, 221, 306
Yorkshire, UK, 34, 44, 46, 105, 106, 229, 230
You can Believe (Musical by Professor Franz Josef Radermacher), 280
You've lost the loving feeling (by the Righteous Brothers), 199

Zoot Money and his Big Roll Band, 28
ZX81,Sinclair, 164